AF725596

Ready For 1st Grade

Math

A Summer Learning Activity Workbook For Students Entering Grade 1 [Ages 6-7]

Copyright © 2026 by Mometrix Media LLC

All rights reserved. This product, or parts thereof, may not be reproduced, stored in a retrieval system, or transmitted in any form or by any means—electronic, mechanical, photocopy, recording, scanning, or other—except for brief quotations in critical reviews or articles, without the prior written permission of the publisher.

Written and edited by Matthew Bowling

Printed in the United States of America

This paper meets the requirements of ANSI/NISO Z39.48-1992 (Permanence of Paper).

Mometrix offers volume discount pricing to institutions. For more information or a price quote, please contact our sales department at sales@mometrix.com or 888-248-1219.

Mometrix Media LLC is not affiliated with or endorsed by any official testing organization. All organizational and test names are trademarks of their respective owners.

Paperback
ISBN 13: 978-1-5167-3115-2
ISBN 10: 1-5167-3115-8

Table of Contents

Introduction

You did it! You finished kindergarten! That is so exciting and awesome!

You are about to head into 1st grade now, and that may be a little scary. This workbook is here to help you feel ready and confident as you tackle this new grade level. You've got this!

This workbook is organized into chapters. Each chapter will review important kindergarten skills and introduce new ideas you'll see in 1st grade. Each chapter starts with a short lesson to refresh your memory or even introduce new things before you take on the activities for each day. You can always go back to that page for help when doing an activity page.

The workbook is made to be completed over 10 weeks, with each day taking no more than 15 minutes. That is just enough time to practice without feeling overwhelmed. Learning is an adventure! As you work through the chapters, the activities will slowly become trickier, but that just means your brain is growing stronger! This will help prepare you for bigger thinking and new tasks you will see in 1st grade.

You might work with a grown-up sometimes, or you might complete much of it on your own. Either way, you're building skills, confidence, and a love for learning. Get ready, because 1st grade is going to be a great year!

Copyright © Mometrix Media. You have been licensed one copy of this document for personal use only. Any other reproduction or redistribution is strictly prohibited. All rights reserved.
This content is provided for test preparation purposes only and does not imply an endorsement by Mometrix of any particular political, scientific, or religious point of view.

Chapter 1: Counting

Counting is a process that determines the number or amount of objects in a set. **Numbers** are standard unique symbols that represent quantity. To count, one must rely on numbers and the understanding of the number sequence. The **number sequence** refers to the order in which numbers are arranged. **Number lines** are visuals that are used to display the correct number sequence on a line. These lines go on forever in both directions, so arrows are placed on either end of the line. Number lines can be displayed vertically or horizontally as shown below and on the side.

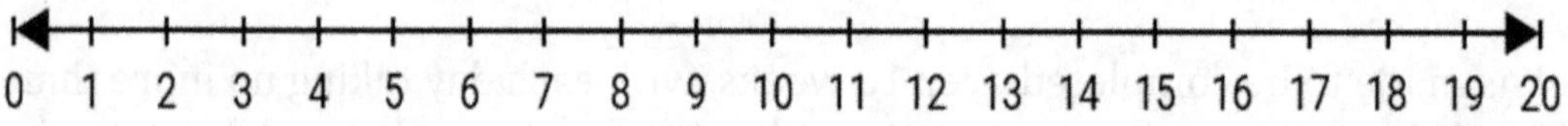

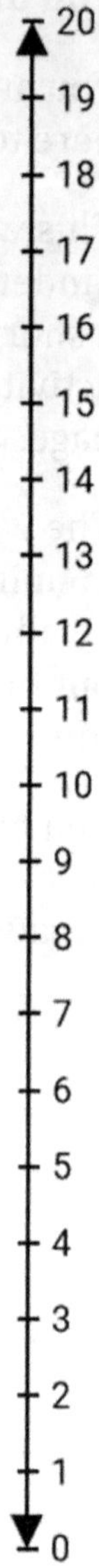

When counting, it is important to remember that each object in a set must be counted exactly once, and the sequence must always stay the same. The last number spoken while counting objects represents how many total items are in the set. Also, the order or direction used to count the objects does not change the total number of items in the set. If the set hasn't changed, it does not need to be recounted each time. If there are no objects in a set, that means there are zero objects. Zero simply means none.

It is also helpful to count in larger increments. This is referred to as **skip counting**. Counting by twos, fives, or tens are all examples of skip counting. An example of counting by fives would be 5, 10, 15, 20, 25, 30, and so on.

It is also helpful to know what is one more or one less of each number. For example, one more than 7 is 8 and one less than 7 is 6.

In Kindergarten, you learned to count to 100 without mistake or assistance. However, in 1st grade, you will learn to count to 120 without mistake or assistance. Keep in mind that numbers follow a pattern, and they are all based on groups of 10. For example, 80 is 8 groups of 10 and 120 is 12 groups of 10.

Complete one worksheet each day in this chapter. Each day should take no longer than 15 minutes to complete.

Copyright © Mometrix Media. You have been licensed one copy of this document for personal use only. Any other reproduction or redistribution is strictly prohibited. All rights reserved.
This content is provided for test preparation purposes only and does not imply an endorsement by Mometrix of any particular political, scientific, or religious point of view.

DAY 1

COUNTING TO 10

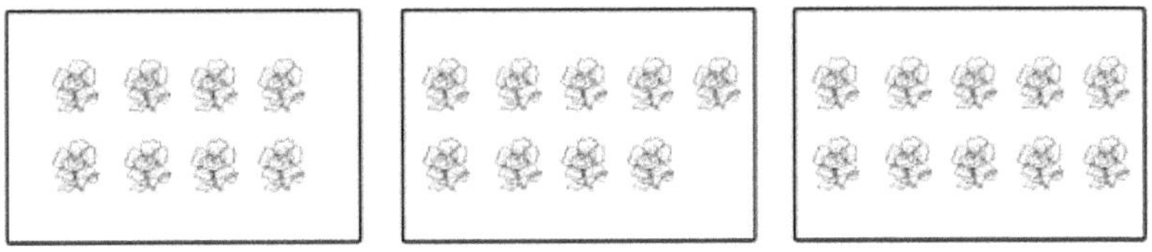

Eight	Nine	Ten
8	9	10

Check the box of the correct number shown in the picture.

1.

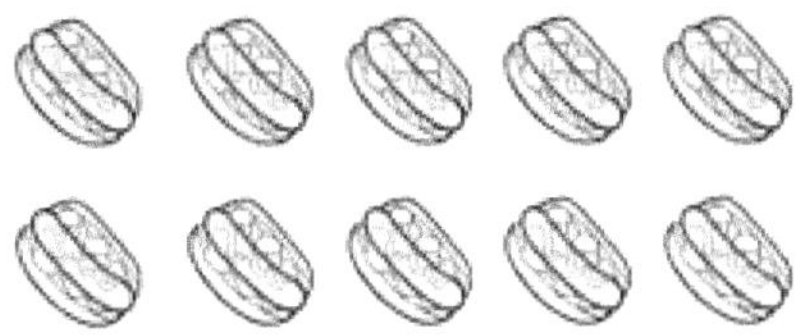

8 ☐ 9 ☐ 10 ☐

2.

8 ☐ 9 ☐ 10 ☐

3.

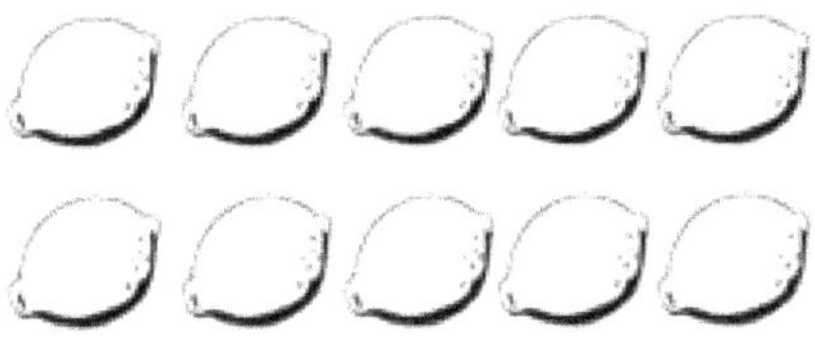

8 ☐ 9 ☐ 10 ☐

4.

8 ☐ 9 ☐ 10 ☐

5.

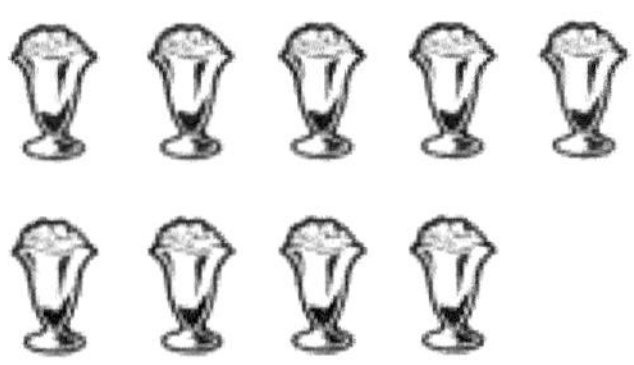

8 ☐ 9 ☐ 10 ☐

6.

8 ☐ 9 ☐ 10 ☐

Copyright © Mometrix Media. You have been licensed one copy of this document for personal use only. Any other reproduction or redistribution is strictly prohibited. All rights reserved.
This content is provided for test preparation purposes only and does not imply an endorsement by Mometrix of any particular political, scientific, or religious point of view.

Day 2

Color the Objects to 10

Color the groups of objects to match the numbers.

1. **4** 
2. **7** 
3. **2** 
4. **6** 
5. **1** 
6. **9** 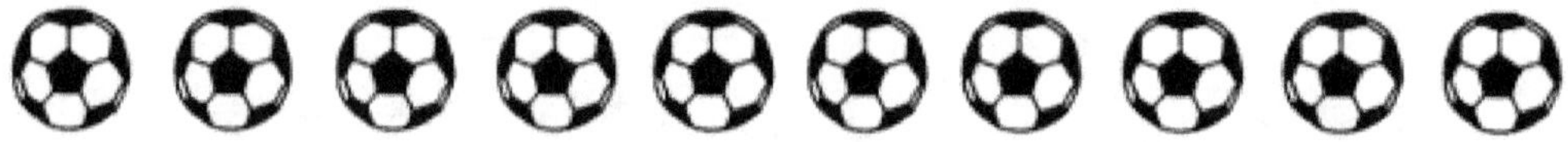
7. **5** 
8. **10** 
9. **8** 
10. **3** 
11. **0** 

Copyright © Mometrix Media. You have been licensed one copy of this document for personal use only. Any other reproduction or redistribution is strictly prohibited. All rights reserved.
This content is provided for test preparation purposes only and does not imply an endorsement by Mometrix of any particular political, scientific, or religious point of view.

DAY 3

COUNTING AND WRITING NUMBERS 1-10 PART 1

Count the objects. Write the number and the word for that number.

3

three

1.

2.

3.

4.

Copyright © Mometrix Media. You have been licensed one copy of this document for personal use only. Any other reproduction or redistribution is strictly prohibited. All rights reserved.
This content is provided for test preparation purposes only and does not imply an endorsement by Mometrix of any particular political, scientific, or religious point of view.

Counting and Writing Numbers 1-10 Part 2

Count the objects. Write the number and the word for that number.

6

six

1.

2.

3.

4.

Copyright © Mometrix Media. You have been licensed one copy of this document for personal use only. Any other reproduction or redistribution is strictly prohibited. All rights reserved.
This content is provided for test preparation purposes only and does not imply an endorsement by Mometrix of any particular political, scientific, or religious point of view.

Day 4

Counting by 2s

Skip count by twos. Write the number for how many in all.

2	4	6		

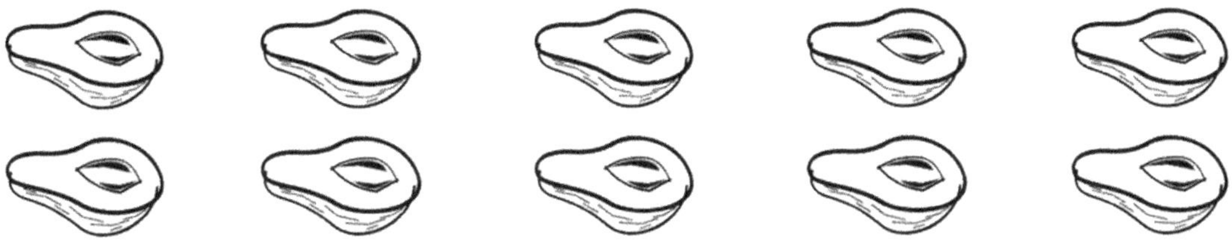

2				

2				

Copyright © Mometrix Media. You have been licensed one copy of this document for personal use only. Any other reproduction or redistribution is strictly prohibited. All rights reserved.
This content is provided for test preparation purposes only and does not imply an endorsement by Mometrix of any particular political, scientific, or religious point of view.

Day 5

Counting 1-10

Count forward. Write the correct number on each object.

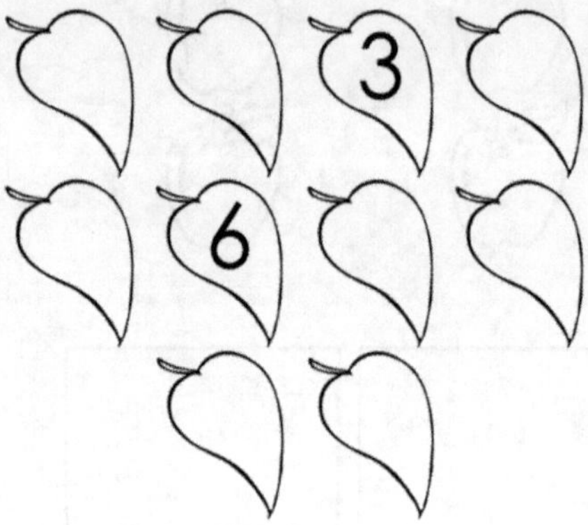

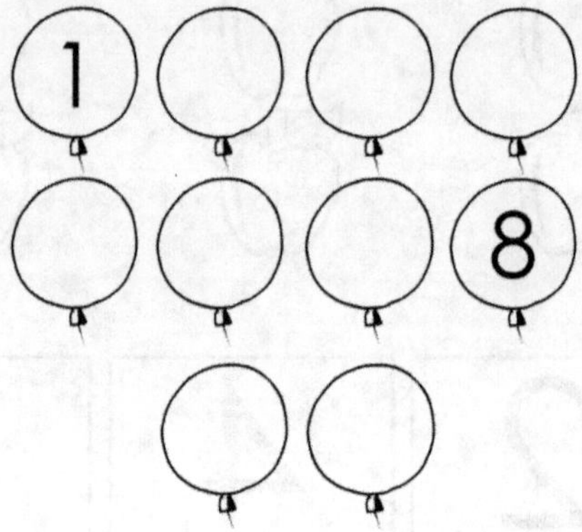

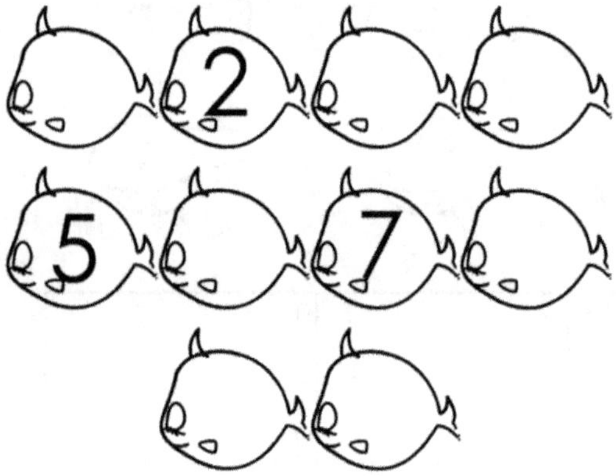

Copyright © Mometrix Media. You have been licensed one copy of this document for personal use only. Any other reproduction or redistribution is strictly prohibited. All rights reserved.
This content is provided for test preparation purposes only and does not imply an endorsement by Mometrix of any particular political, scientific, or religious point of view.

Day 6

Counting Objects 1-20

Count the objects. Write the correct number for each set in the box.

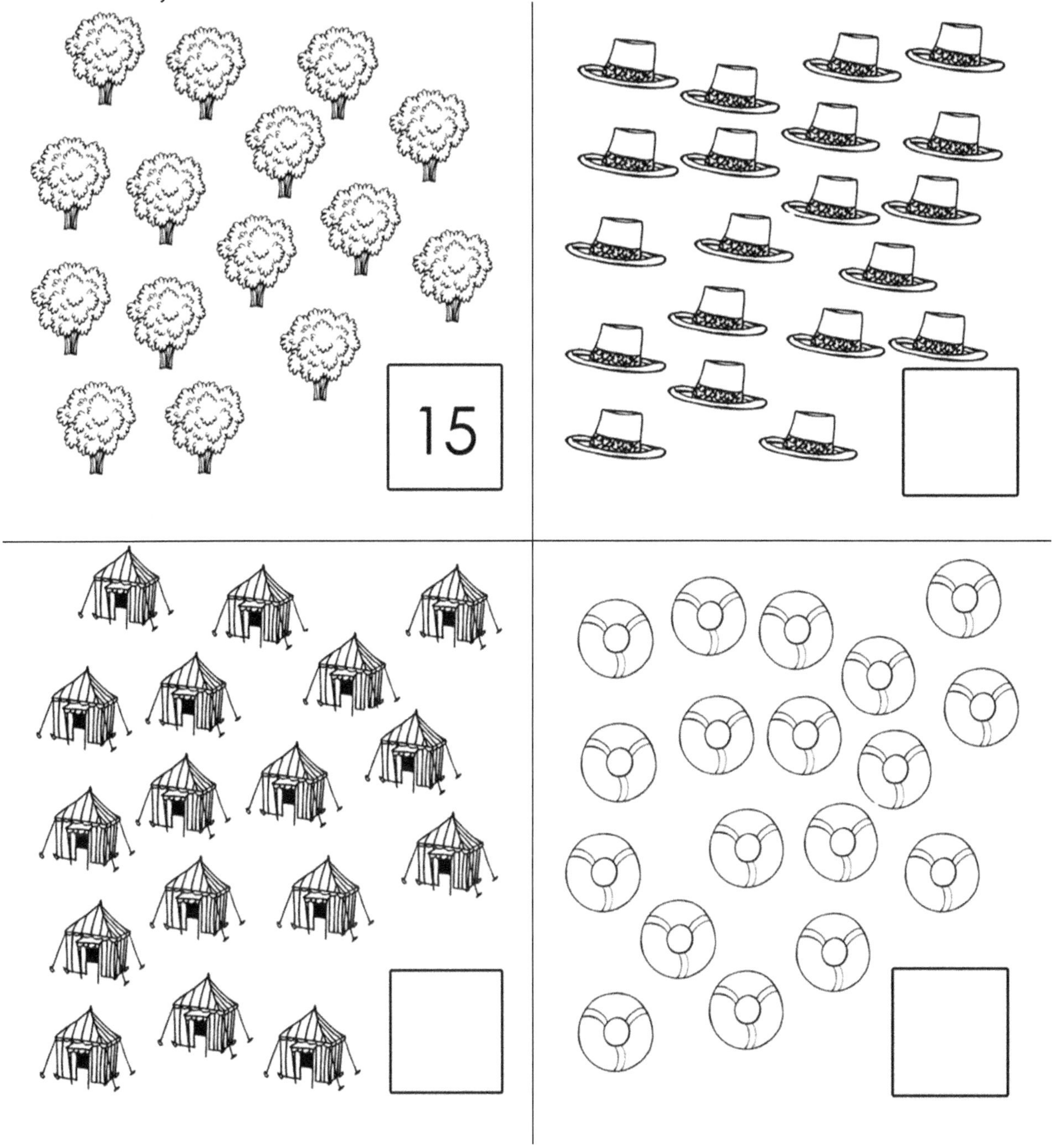

Copyright © Mometrix Media. You have been licensed one copy of this document for personal use only. Any other reproduction or redistribution is strictly prohibited. All rights reserved.
This content is provided for test preparation purposes only and does not imply an endorsement by Mometrix of any particular political, scientific, or religious point of view.

Day 7

Counting Objects 1-50

Count the objects. Write the correct number for each set in the box.

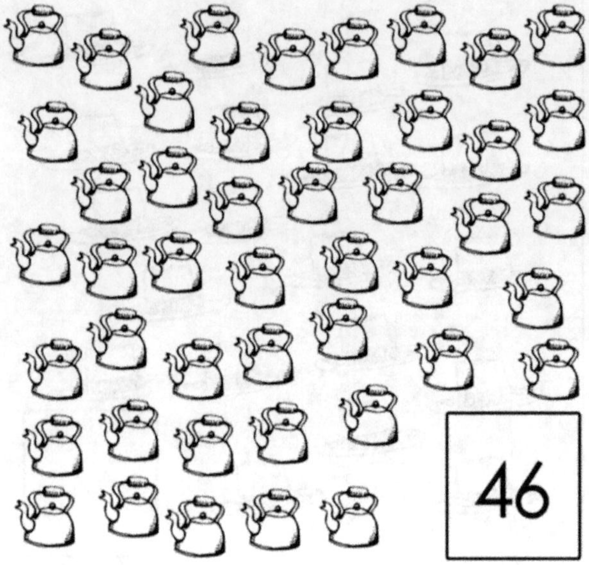

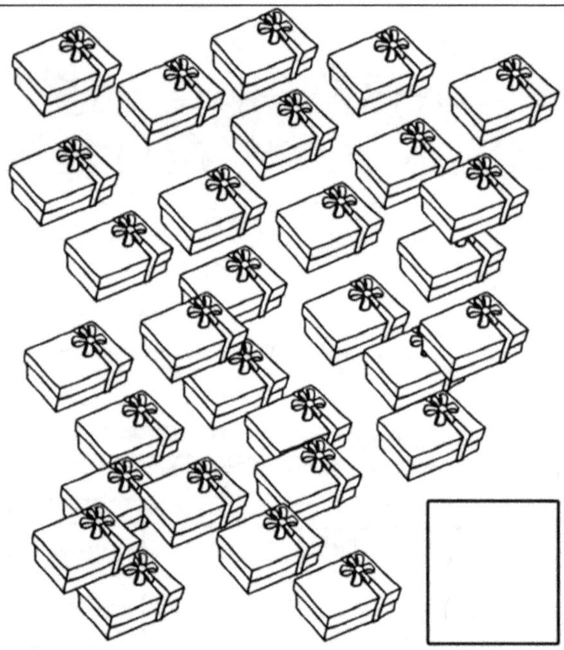

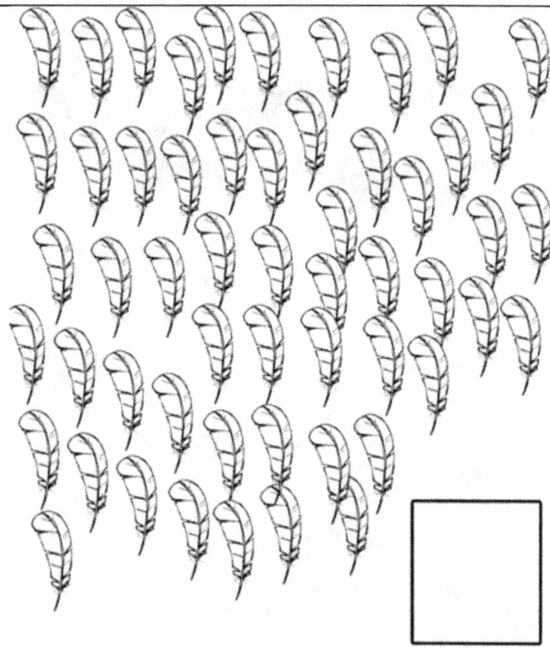

Copyright © Mometrix Media. You have been licensed one copy of this document for personal use only. Any other reproduction or redistribution is strictly prohibited. All rights reserved.
This content is provided for test preparation purposes only and does not imply an endorsement by Mometrix of any particular political, scientific, or religious point of view.

Day 8

Counting 1-120

Fill in the missing number on the chart below. Use this chart to help you practice counting to 120.

1	2		4			7		9	
11		13		15	16		18		
21			24		26			29	30
31	32			35			38		40
	42	43	44				48		
51		53		55	56		58		60
	62	63			66	67		69	
71			74	75					80
	82		84		86	87		89	
		93	94	95					100
101						107			
		113						119	

Copyright © Mometrix Media. You have been licensed one copy of this document for personal use only. Any other reproduction or redistribution is strictly prohibited. All rights reserved.
This content is provided for test preparation purposes only and does not imply an endorsement by Mometrix of any particular political, scientific, or religious point of view.

Chapter 2: Addition

Addition is when two or more sets of things are combined or joined together. This is also called **composing**. The **sum** is the answer found after adding the sets. Similar to counting, unique symbols are used to indicate addition. The **plus sign** (+) represents addition, and the **equal sign** (=) is used to indicate the sum or answer. Numbers are used to represent the sets being combined.

Addition can be done using numbers, visuals, and physical objects.

Example:

Sarah drew 5 circles on her paper. Tom drew 4 circles on his paper. How many circles are there all together?

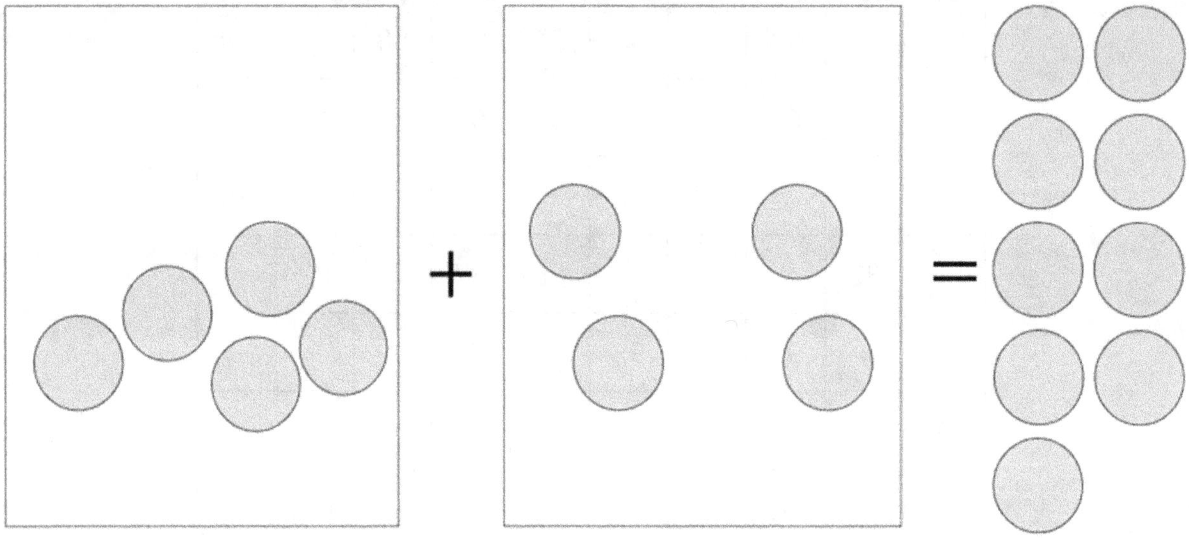

$$5 + 4 = 9$$

Place value plays an important role in addition and adding larger numbers. It is important to add values in each place value position with other values in the same place value position (i.e., ones with ones, tens with tens, etc.). When the sum of each place value is greater than 9, an additional group is added to the next place value position.

Complete one worksheet each day in this chapter. Each day should take no longer than 15 minutes to complete.

Copyright © Mometrix Media. You have been licensed one copy of this document for personal use only. Any other reproduction or redistribution is strictly prohibited. All rights reserved.
This content is provided for test preparation purposes only and does not imply an endorsement by Mometrix of any particular political, scientific, or religious point of view.

Day 9

Addition to 5

Add the objects in each row. Write the total in the box.

 +  = 5

 + 0 = ☐

 +  +  = ☐

 +  +  +  = ☐

 +  = ☐

Copyright © Mometrix Media. You have been licensed one copy of this document for personal use only. Any other reproduction or redistribution is strictly prohibited. All rights reserved.
This content is provided for test preparation purposes only and does not imply an endorsement by Mometrix of any particular political, scientific, or religious point of view.

DAY 10

ADDITION GROUPS 1-10

Find the sum of each.

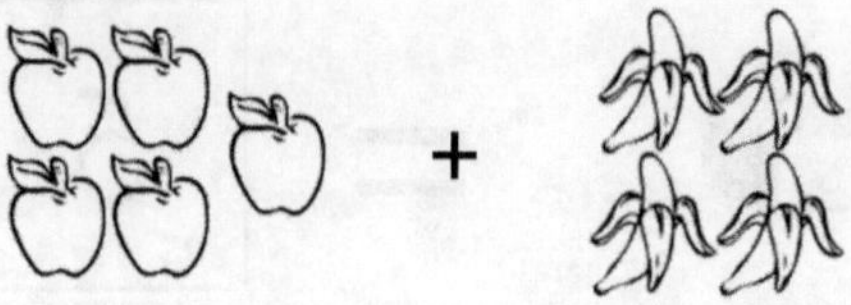

5
+ 4
9

2
+ 2

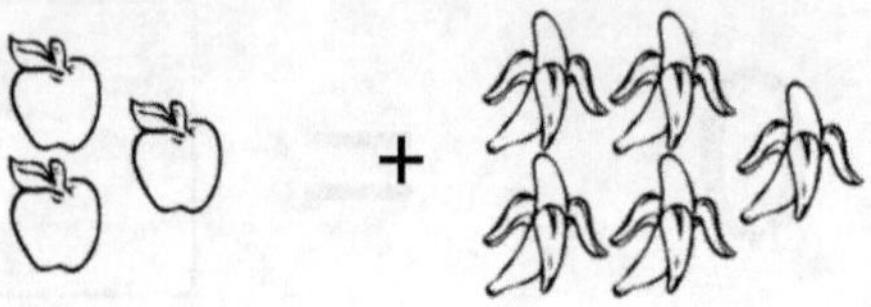

3
+ 5

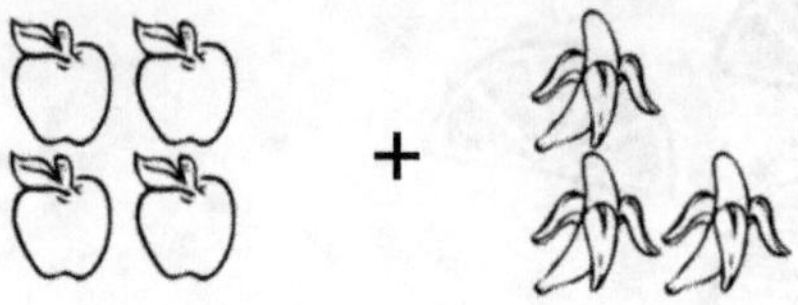

4
+ 3

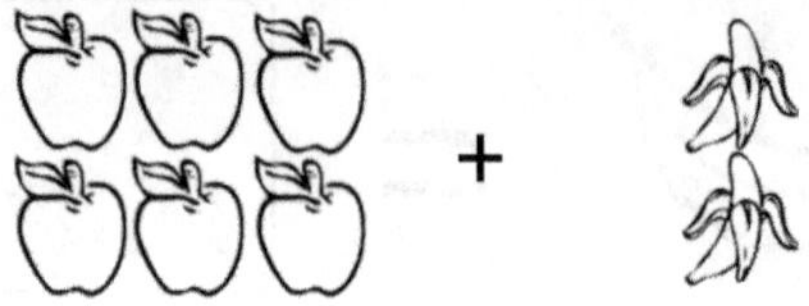

6
+ 2

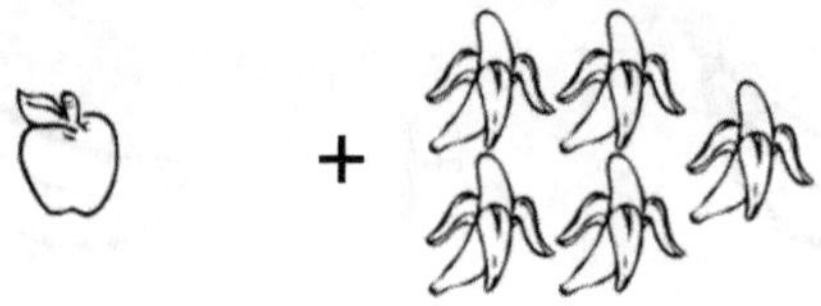

1
+ 5

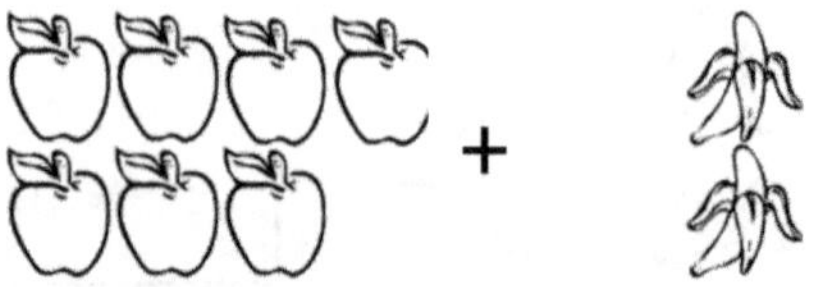

7
+ 2

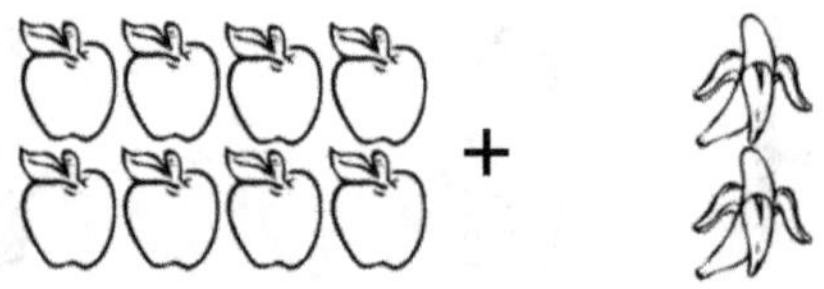

8
+ 2

Copyright © Mometrix Media. You have been licensed one copy of this document for personal use only. Any other reproduction or redistribution is strictly prohibited. All rights reserved.
This content is provided for test preparation purposes only and does not imply an endorsement by Mometrix of any particular political, scientific, or religious point of view.

Day 11

Word Problems 1-10

Solve each problem. Show your work.

We see 5 

1 more  shows up

What is the sum? 6

$$\begin{array}{r} 5 \\ +\ 1 \\ \hline 6 \end{array}$$

There is 1 

9 more  arrive

What is the total? _____

There are 4 

4 more  show up

What is the sum? _____

We have 3

1 swims up

How many are there?

___

There are 2 

7 more  arrive

What is the total? _____

There are 3 

5 more  walk up

What is the sum? _____

Jimmy has 2 

Sue has 4 

What is the sum? _____

Steve has 5 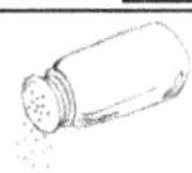

He added 2 

What is the total? _____

Copyright © Mometrix Media. You have been licensed one copy of this document for personal use only. Any other reproduction or redistribution is strictly prohibited. All rights reserved.
This content is provided for test preparation purposes only and does not imply an endorsement by Mometrix of any particular political, scientific, or religious point of view.

Day 12

Addition 1-10

Add the groups. Write the total.

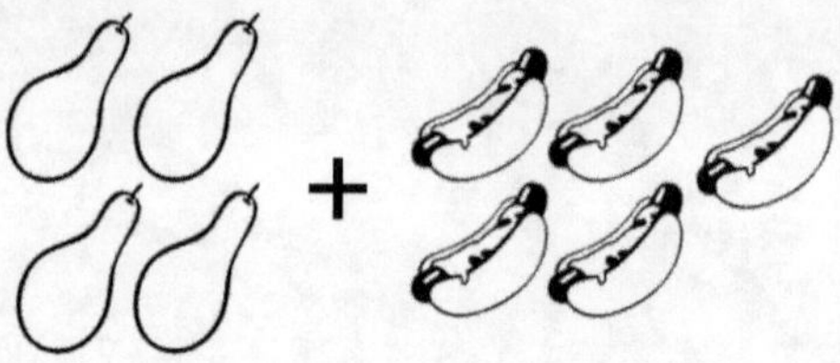

4
+ 5
9

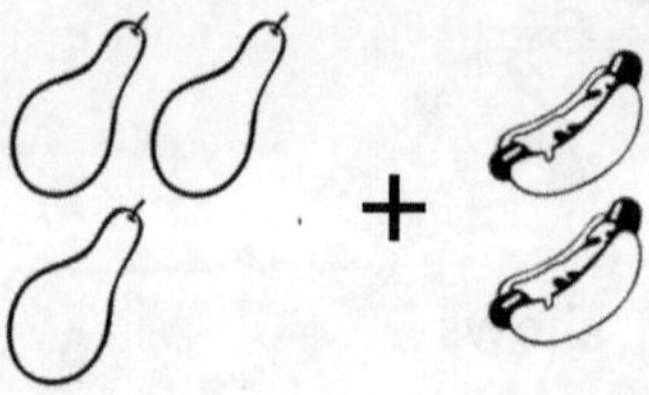

3
+ 2

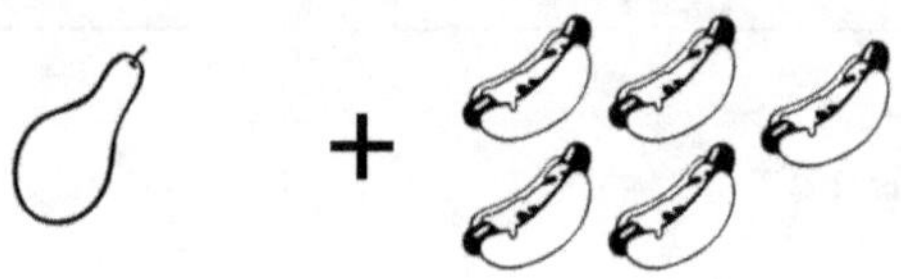

1
+ 5

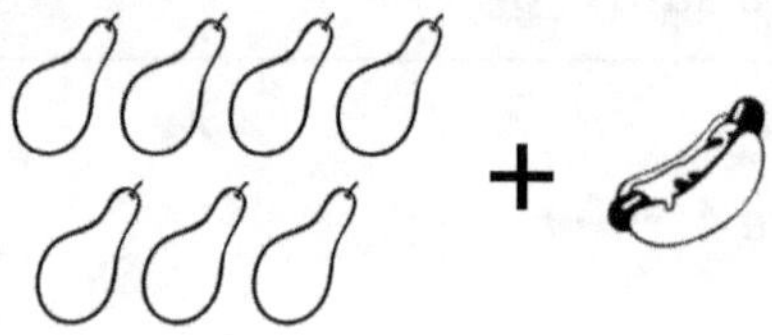

7
+ 1

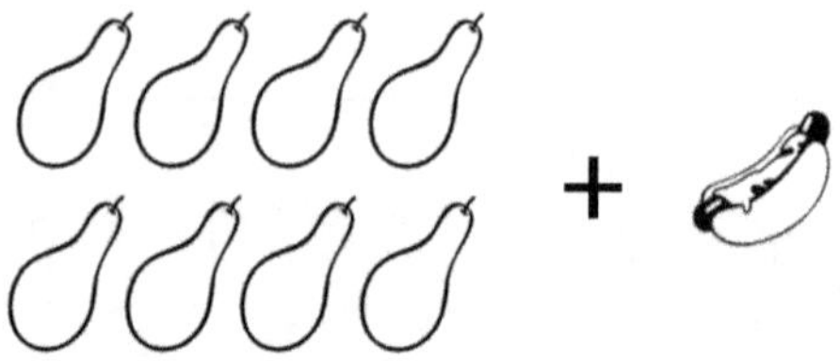

8
+ 1

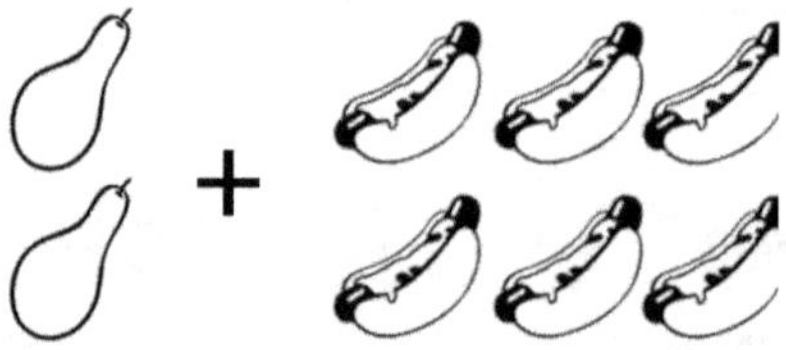

2
+ 6

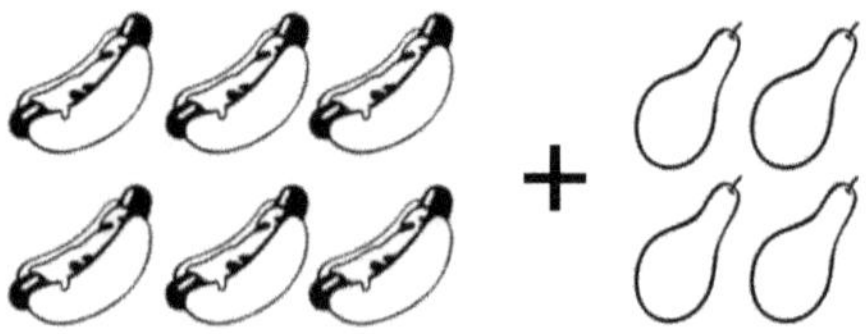

6
+ 4

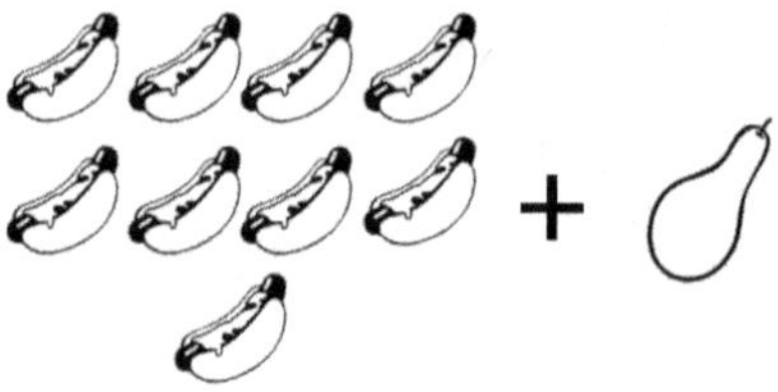

9
+ 1

Copyright © Mometrix Media. You have been licensed one copy of this document for personal use only. Any other reproduction or redistribution is strictly prohibited. All rights reserved.
This content is provided for test preparation purposes only and does not imply an endorsement by Mometrix of any particular political, scientific, or religious point of view.

Day 13

Addition 1-12

Add the groups. Write the total.

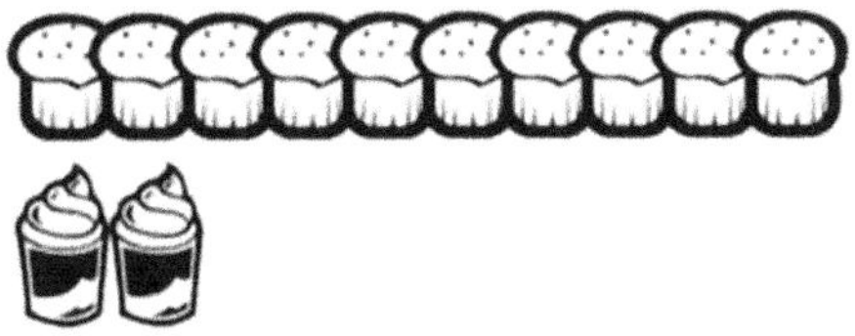

$$\begin{array}{r} 10 \\ +\ 2 \\ \hline 12 \end{array}$$

$$\begin{array}{r} 6 \\ +\ 4 \\ \hline \end{array}$$

$$\begin{array}{r} 8 \\ +\ 3 \\ \hline \end{array}$$

$$\begin{array}{r} 5 \\ +\ 4 \\ \hline \end{array}$$

$$\begin{array}{r} 3 \\ +\ 9 \\ \hline \end{array}$$

$$\begin{array}{r} 4 \\ +\ 4 \\ \hline \end{array}$$

$$\begin{array}{r} 6 \\ +\ 5 \\ \hline \end{array}$$

$$\begin{array}{r} 9 \\ +\ 3 \\ \hline \end{array}$$

Copyright © Mometrix Media. You have been licensed one copy of this document for personal use only. Any other reproduction or redistribution is strictly prohibited. All rights reserved.
This content is provided for test preparation purposes only and does not imply an endorsement by Mometrix of any particular political, scientific, or religious point of view.

Day 14

Missing Numbers Addition 1-10

Write the missing numbers in the boxes.

4 + [2] = 6	[ ] + 3 = 5
2 + [ ] = 9	4 + [ ] = 4
[ ] + 2 = 5	6 + [ ] = 7
5 + [ ] = 9	[ ] + 3 = 10
[ ] + 8 = 10	2 + [ ] = 8

Copyright © Mometrix Media. You have been licensed one copy of this document for personal use only. Any other reproduction or redistribution is strictly prohibited. All rights reserved.
This content is provided for test preparation purposes only and does not imply an endorsement by Mometrix of any particular political, scientific, or religious point of view.

DAY 15

MISSING NUMBERS ADDITION 1-20

Write the missing numbers in the boxes.

5 + [11] = 16	[ ] + 5 = 19
4 + [ ] = 11	12 + [ ] = 17
[ ] + 2 = 15	6 + [ ] = 20
5 + [ ] = 19	[ ] + 3 = 13
[ ] + 5 = 20	2 + [ ] = 18

Copyright © Mometrix Media. You have been licensed one copy of this document for personal use only. Any other reproduction or redistribution is strictly prohibited. All rights reserved.
This content is provided for test preparation purposes only and does not imply an endorsement by Mometrix of any particular political, scientific, or religious point of view.

Day 16

Addition 1-20

Solve the problems below by adding.

1. 10 + 9 = 19
2. 2 + 7 = ____
3. 10 + 7 = ____
4. 2 + 7 = ____
5. 10 + 6 = ____
6. 10 + 1 = ____
7. 10 + 5 = ____
8. 1 + 2 = ____
9. 9 + 3 = ____
10. 8 + 9 = ____
11. 19 + 1 = ____
12. 4 + 6 = ____
13. 8 + 9 = ____
14. 1 + 19 = ____
15. 13 + 7 = ____
16. 1 + 8 = ____
17. 10 + 10 = ____
18. 16 + 3 = ____
19. 1 + 13 = ____
20. 11 + 9 = ____
21. 2 + 5 = ____
22. 7 + 9 = ____
23. 8 + 8 = ____
24. 1 + 11 = ____
25. 3 + 7 = ____
26. 9 + 5 = ____
27. 3 + 5 = ____
28. 14 + 6 = ____
29. 18 + 1 = ____
30. 3 + 9 = ____

Copyright © Mometrix Media. You have been licensed one copy of this document for personal use only. Any other reproduction or redistribution is strictly prohibited. All rights reserved.
This content is provided for test preparation purposes only and does not imply an endorsement by Mometrix of any particular political, scientific, or religious point of view.

Chapter 3: Subtraction

Subtraction is when a set of things has a portion that is separated, removed, or taken away. This is also referred to as **decomposing**. The **difference** is the answer found after subtracting from the set. Similar to counting and addition, unique symbols are used to indicate subtraction. The **minus sign** (-) represents subtraction, and the **equal sign** (=) is used to indicate the difference or answer. Numbers are used to represent the amounts being subtracted.

Subtraction can be done using numbers, visuals, and physical objects.

Example:

James drew 6 shapes on his paper. He then cut out 2 of the shapes. How many shapes does James have left to cut out?

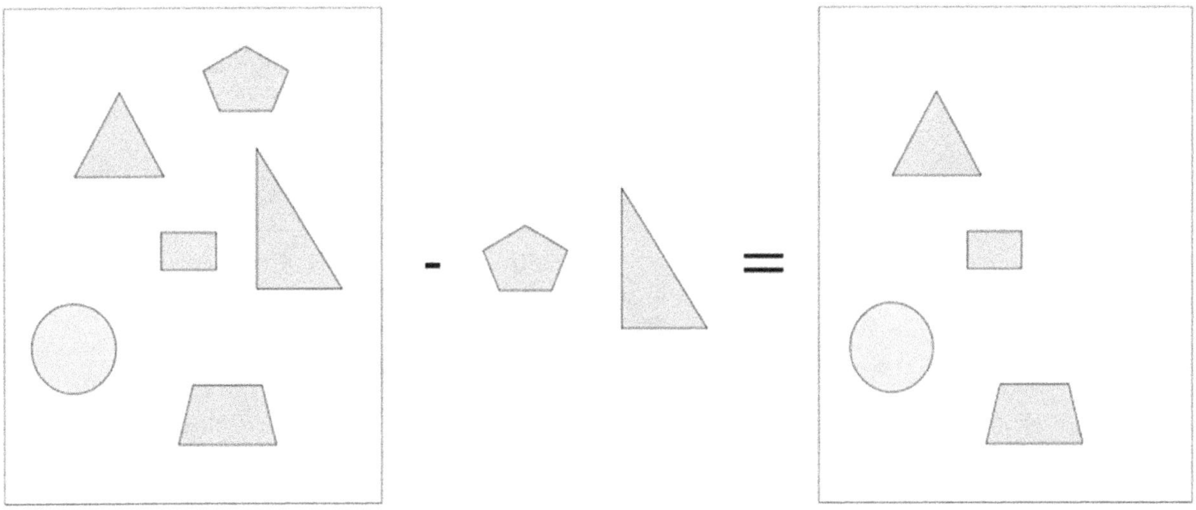

$$6 - 2 = 4$$

Place value plays an important role in subtraction and subtracting larger numbers. It is important to subtract values in each place value position with other values in the same place value position (i.e. ones with ones, tens with tens, etc.). When the number being subtracted from has a smaller digit value than the number that is being subtracted, a group is borrowed from the next place value position.

Complete one worksheet each day in this chapter. Each day should take no longer than 15 minutes to complete.

Copyright © Mometrix Media. You have been licensed one copy of this document for personal use only. Any other reproduction or redistribution is strictly prohibited. All rights reserved.
This content is provided for test preparation purposes only and does not imply an endorsement by Mometrix of any particular political, scientific, or religious point of view.

Day 17

Subtraction to 5

Subtract the objects in each row. Write the difference in the box.

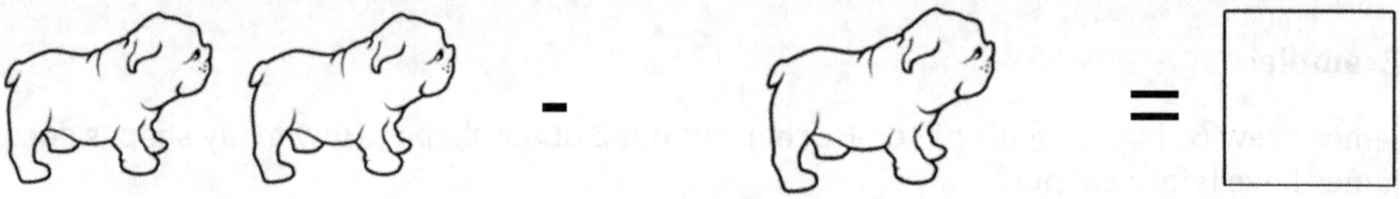

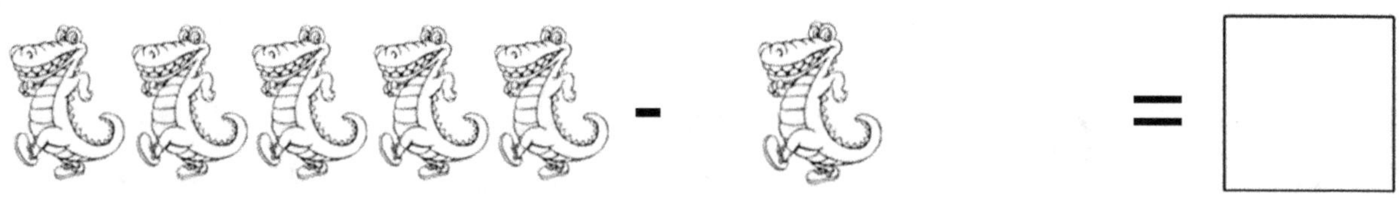

Copyright © Mometrix Media. You have been licensed one copy of this document for personal use only. Any other reproduction or redistribution is strictly prohibited. All rights reserved.
This content is provided for test preparation purposes only and does not imply an endorsement by Mometrix of any particular political, scientific, or religious point of view.

DAY 18

WORD PROBLEMS

Solve each problem. Show your work.

We see 4 

2  run away

How many are left? 2

$$\begin{array}{r} 4 \\ -\ 2 \\ \hline 2 \end{array}$$

There are 6 

2  go home

How many are left? ______

There are 6 

1  drives away

How many are left? ______

We have 4 

1  leaves

How many are left? ______

There are 7 

3  leave

How many are left? ______

There are 5 

2  drive away

What is the difference?

______

Jimmy has 2 

Sue has 1 

What is the difference? ______

Steve has 5 

He ate 1 

How many are left? ______

Chapter 3: Subtraction

Copyright © Mometrix Media. You have been licensed one copy of this document for personal use only. Any other reproduction or redistribution is strictly prohibited. All rights reserved.
This content is provided for test preparation purposes only and does not imply an endorsement by Mometrix of any particular political, scientific, or religious point of view.

Day 19

Subtraction to 10

Solve each problem. Show your work.

5
- 4
_____
1

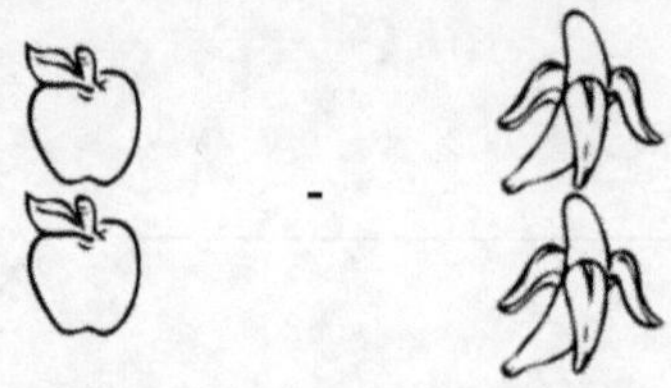

2
- 2
_____

5
- 3
_____

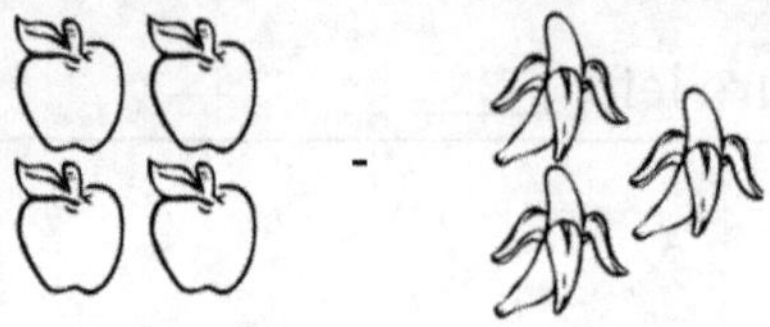

4
- 3
_____

6
- 2
_____

1
- 1
_____

7
- 2
_____

8
- 2
_____

Copyright © Mometrix Media. You have been licensed one copy of this document for personal use only. Any other reproduction or redistribution is strictly prohibited. All rights reserved.
This content is provided for test preparation purposes only and does not imply an endorsement by Mometrix of any particular political, scientific, or religious point of view.

Day 20

Subtraction to 12

Solve each problem.

12
\- 2
———
10

10
\- 4
———

11
\- 3
———

9
\- 4
———

5
\- 4
———

7
\- 3
———

10
\- 5
———

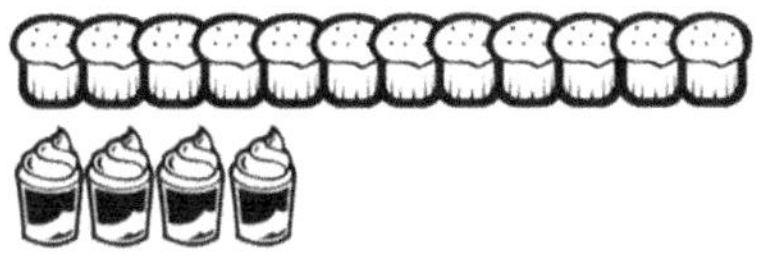

12
\- 4
———

Copyright © Mometrix Media. You have been licensed one copy of this document for personal use only. Any other reproduction or redistribution is strictly prohibited. All rights reserved.
This content is provided for test preparation purposes only and does not imply an endorsement by Mometrix of any particular political, scientific, or religious point of view.

Day 21
Missing Numbers to 10

Write the missing numbers in the boxes.

6	-	4	=	2
8	-		=	4
	-	2	=	6
5	-		=	1
	-	2	=	8

	-	3	=	5
7	-		=	3
9	-		=	7
	-	3	=	10
10	-		=	8

Copyright © Mometrix Media. You have been licensed one copy of this document for personal use only. Any other reproduction or redistribution is strictly prohibited. All rights reserved.
This content is provided for test preparation purposes only and does not imply an endorsement by Mometrix of any particular political, scientific, or religious point of view.

Day 22
Missing Numbers to 20

Write the missing numbers in the boxes.

18	-	2	=	16
17	-	☐	=	4
☐	-	2	=	14
15	-	☐	=	1
☐	-	2	=	17

☐	-	1	=	5
12	-	☐	=	2
9	-	☐	=	7
☐	-	3	=	17
16	-	☐	=	8

Copyright © Mometrix Media. You have been licensed one copy of this document for personal use only. Any other reproduction or redistribution is strictly prohibited. All rights reserved.
This content is provided for test preparation purposes only and does not imply an endorsement by Mometrix of any particular political, scientific, or religious point of view.

Day 23

Subtraction Problems to 15

Solve the problems.

1.	2.	3.	4.	5.	6.	7.
10	7	5	8	3	10	10
- 9	- 2	- 3	- 4	- 2	- 1	- 10
1						

8.	9.	10.	11.	12.	13.	14.
7	5	9	5	6	8	1
- 1	- 3	- 8	- 1	- 4	- 2	- 1

15.	16.	17.	18.	19.	20.	21.
13	8	10	6	7	10	5
- 7	- 1	- 5	- 3	- 2	- 7	- 2

22.	23.	24.	25.	26.	27.	28.
8	8	10	7	9	5	14
- 7	- 8	- 1	- 3	- 5	- 3	- 6

29.	30.
4	9
- 1	- 3

Copyright © Mometrix Media. You have been licensed one copy of this document for personal use only. Any other reproduction or redistribution is strictly prohibited. All rights reserved.
This content is provided for test preparation purposes only and does not imply an endorsement by Mometrix of any particular political, scientific, or religious point of view.

Day 24

Subtraction Problems to 20

Solve the problems.

1. 19 - 9 = 10
2. 17 - 2
3. 15 - 3
4. 18 - 4
5. 11 - 2
6. 10 - 1
7. 11 - 10
8. 17 - 1
9. 12 - 3
10. 13 - 8
11. 20 - 1
12. 16 - 4
13. 18 - 9
14. 10 - 1
15. 20 - 7
16. 12 - 8
17. 10 - 10
18. 7 - 3
19. 14 - 2
20. 15 - 9
21. 9 - 5
22. 8 - 7
23. 18 - 8
24. 10 - 1
25. 17 - 3
26. 19 - 5
27. 20 - 3
28. 14 - 6
29. 18 - 5
30. 12 - 3

Copyright © Mometrix Media. You have been licensed one copy of this document for personal use only. Any other reproduction or redistribution is strictly prohibited. All rights reserved.
This content is provided for test preparation purposes only and does not imply an endorsement by Mometrix of any particular political, scientific, or religious point of view.

Chapter 4: Place Value, Sorting & Number Sense

Numbers can be broken apart and composed using **place value**. For example, the number 129,548 can be expressed in **expanded form** as 100,000 + 20,000 + 9,000 + 500 + 40 + 8. You should also be comfortable going from expanded form to standard form. For example, what number is represented by 30,000 + 400 + 80 + 2? Notice 3 groups of ten-thousands, 0 groups of thousands, 4 groups of hundreds, 8 groups of tens, and 2 groups of ones. This becomes 30,482.

Sometimes you will be asked to compose and decompose numbers that are presented in **picture** or **model** form. This usually includes base-10 blocks. For example, consider the model below. How many groups of thousands, hundreds, tens, and ones are there? There is 1 group of 1,000, 2 groups of 100, 4 groups of 10, and 7 ones. This forms the number 1,247.

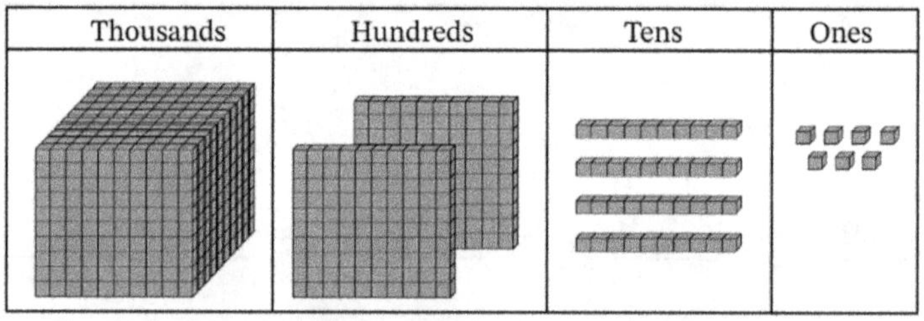

Two important symbols are the **inequality symbols** for **greater than** (>) and **less than** (<). For example, 45 > 35 means 45 is greater than 35, and 26 < 62 means 26 is less than 62. We can now build on this skill to compare and order whole numbers up to 100,000. The important thing to remember is place value. Remember, when comparing numbers, start by comparing the **farthest left digit**. For example, when comparing 42 and 45 start by comparing the tens place.

42 and 45

In this case, both values are 4. Since they are the same, move to the ones place.

42 and 45

2 is less than 5, so 42 is less than 45. This can be expressed using an inequality symbol: 42 < 45.

This also means that 45 is greater than 42. This can be expressed using an inequality symbol: 45 > 42.

Complete one worksheet each day in this chapter. Each day should take no longer than 15 minutes to complete.

Copyright © Mometrix Media. You have been licensed one copy of this document for personal use only. Any other reproduction or redistribution is strictly prohibited. All rights reserved.
This content is provided for test preparation purposes only and does not imply an endorsement by Mometrix of any particular political, scientific, or religious point of view.

Day 25

Sorting

Circle or color the object that does not match in each group.

Copyright © Mometrix Media. You have been licensed one copy of this document for personal use only. Any other reproduction or redistribution is strictly prohibited. All rights reserved.
This content is provided for test preparation purposes only and does not imply an endorsement by Mometrix of any particular political, scientific, or religious point of view.

Day 26

Grouping Addition

Count the groups. Write the totals in the blanks.

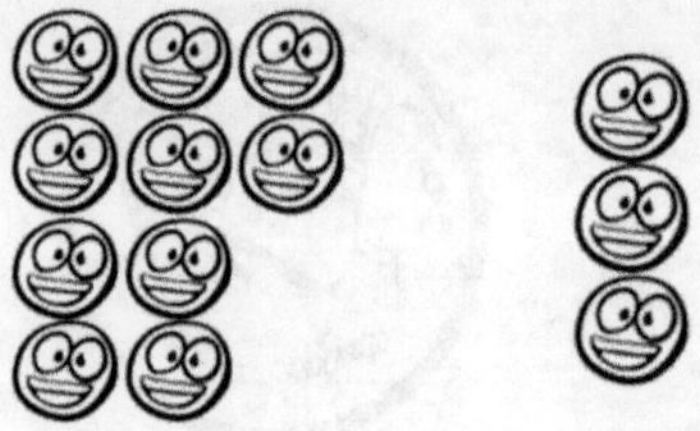

10 + 3 = 13

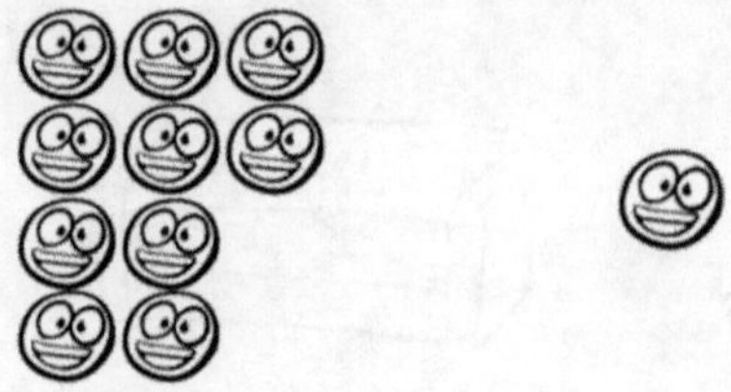

10 + 1 = ____

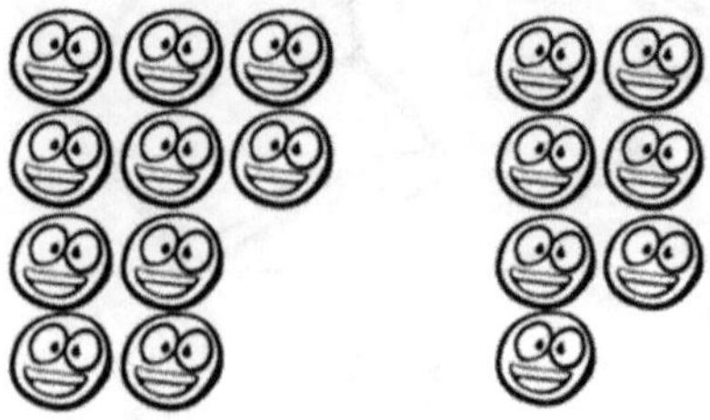

10 + 7 = ____

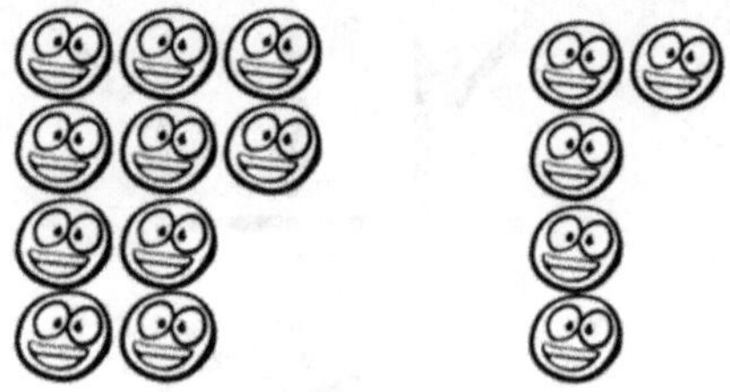

10 + 5 = ____

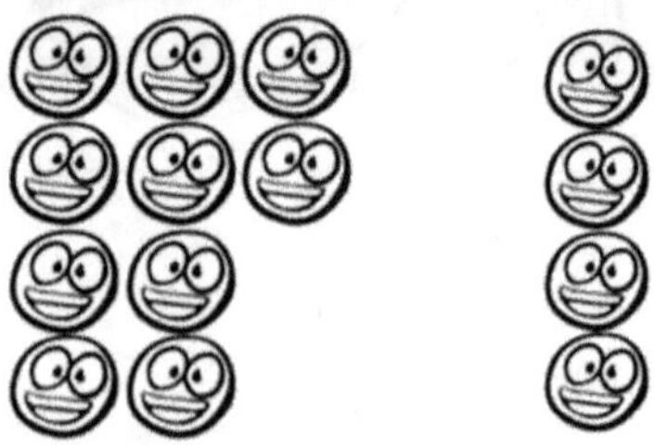

10 + 4 = ____

10 + 9 = ____

Copyright © Mometrix Media. You have been licensed one copy of this document for personal use only. Any other reproduction or redistribution is strictly prohibited. All rights reserved.
This content is provided for test preparation purposes only and does not imply an endorsement by Mometrix of any particular political, scientific, or religious point of view.

Day 27

Grouping Addition

Write out the values and then add them.

 = 10  = 1

 30 + 4 34	 + ___
 + ___	 + ___
 + ___	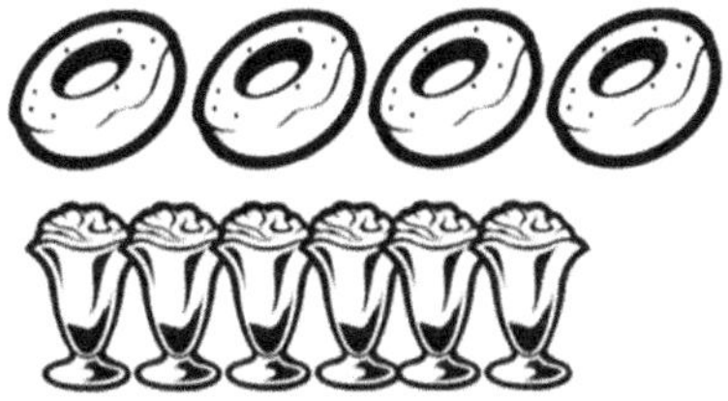 + ___
 + ___	 + ___

Chapter 4: Place Value, Sorting & Number Sense

Copyright © Mometrix Media. You have been licensed one copy of this document for personal use only. Any other reproduction or redistribution is strictly prohibited. All rights reserved.
This content is provided for test preparation purposes only and does not imply an endorsement by Mometrix of any particular political, scientific, or religious point of view.

DAY 28

GROUPING SUBTRACTION

Write out the values and then subtract them.

 = 10  = 1

40
\- 2
38

\-

\-

\-

\-

\-

\-

\-

Copyright © Mometrix Media. You have been licensed one copy of this document for personal use only. Any other reproduction or redistribution is strictly prohibited. All rights reserved.
This content is provided for test preparation purposes only and does not imply an endorsement by Mometrix of any particular political, scientific, or religious point of view.

Day 29

Comparing Groups

Compare the object groups. Draw a > that points to the smaller group.

> = Greater than, < = Less than.

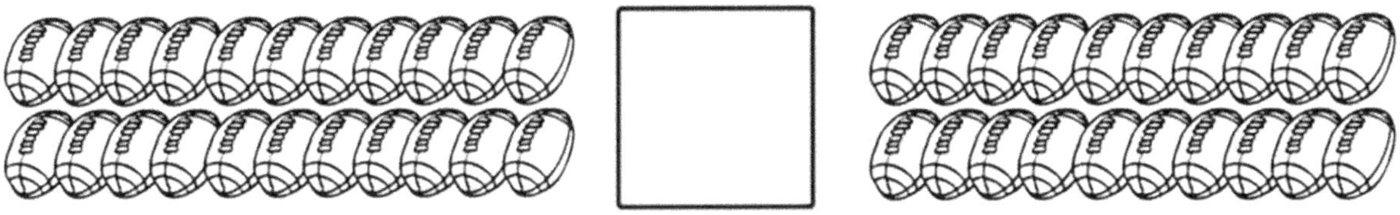

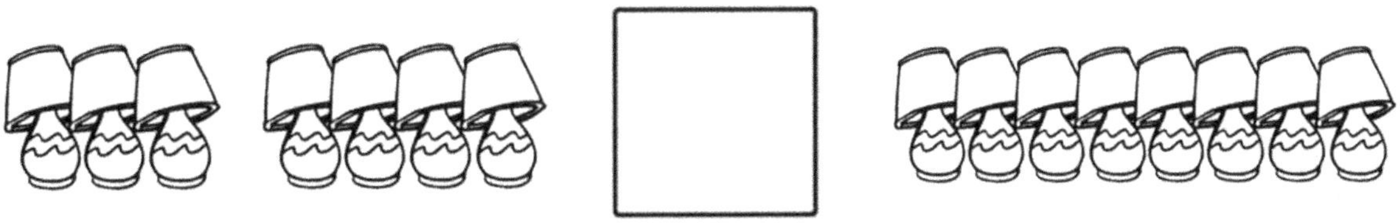

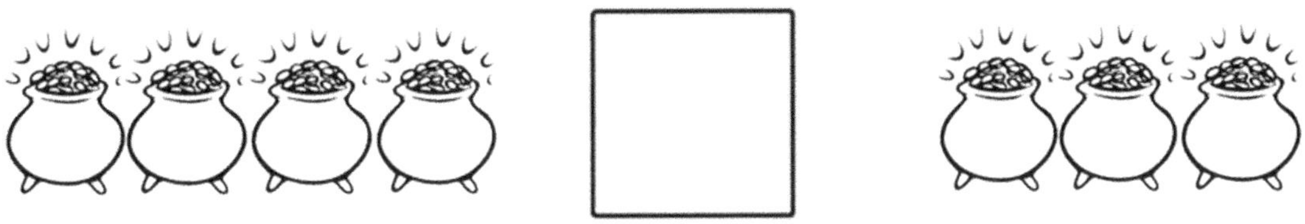

Copyright © Mometrix Media. You have been licensed one copy of this document for personal use only. Any other reproduction or redistribution is strictly prohibited. All rights reserved.
This content is provided for test preparation purposes only and does not imply an endorsement by Mometrix of any particular political, scientific, or religious point of view.

DAY 30

COMPARING GROUPS

Compare the object groups. Draw a > that points to the smaller group.

> = Greater than, < = Less than.

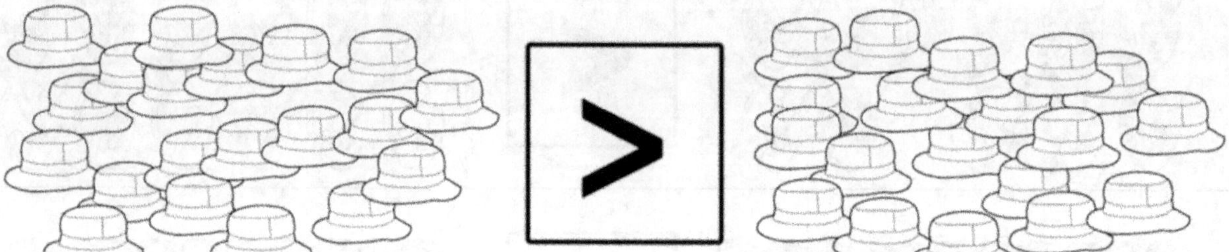

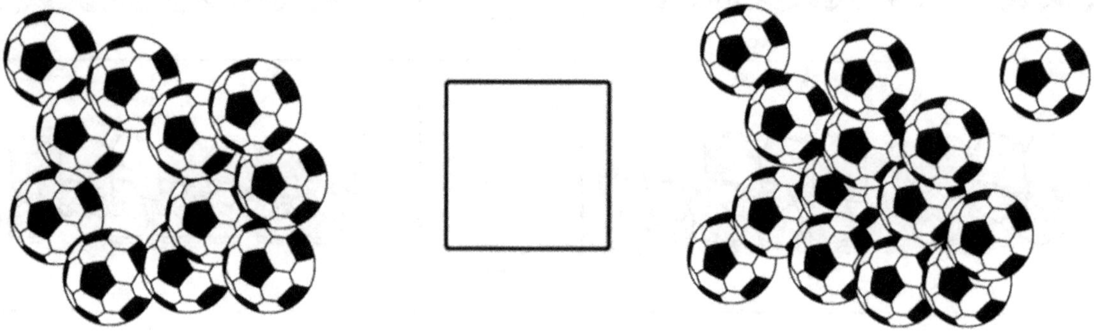

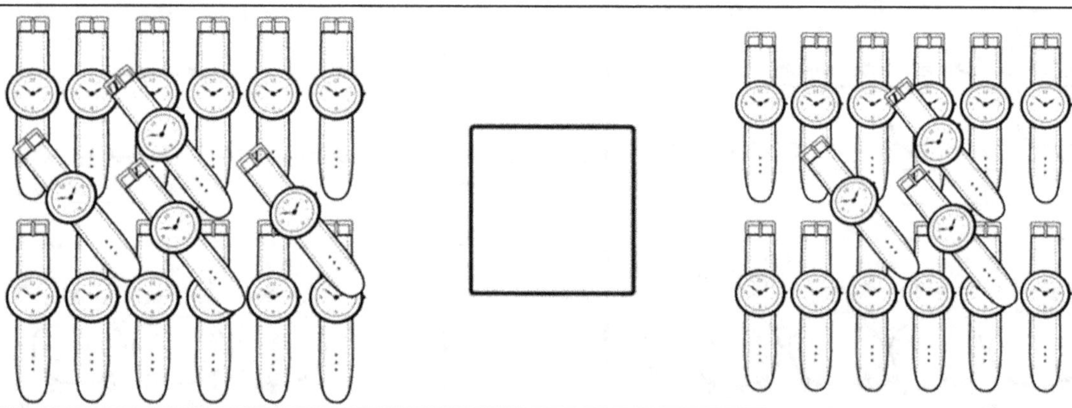

Copyright © Mometrix Media. You have been licensed one copy of this document for personal use only. Any other reproduction or redistribution is strictly prohibited. All rights reserved.
This content is provided for test preparation purposes only and does not imply an endorsement by Mometrix of any particular political, scientific, or religious point of view.

Chapter 5: Time

The clock below displays 4:00. The short hand, referred to as the **hour hand**, points to 4, and the long hand, referred to as the **minute hand**, points to 12, which represents 0 minutes.

Moving to the right around the clock starting at 12 is referred to as **clockwise**. In order to tell how many minutes have passed in each hour, the minute hand moves clockwise around the clock. Each number starting at 1 represents an interval of 5. For example, 1 represents 5 minutes, 2 represents 10 minutes, 3 represents 15 minutes, and so on. When the minute hand reaches 12, the number of minutes resets to zero, and the hour hand moves clockwise to the next number to represent the hour. It is also important to note that the hour hand moves slightly as minutes pass, and it will be closest to the nearest hour mark. For example, if the time is 4:50, the hour hand will be closer to 5 than 4. Likewise, if the time is 4:10, the hour hand will be closer to 4 than 5.

There are a total of 60 minutes in one hour. One-half of one hour is 30 minutes. One-quarter or one-fourth of one hour is 15 minutes. 45 minutes is three-quarters of an hour or one-quarter of an hour till the next hour since there are only 15 minutes left in the current hour.

Complete one worksheet each day in this chapter. Each day should take no longer than 15 minutes to complete.

Copyright © Mometrix Media. You have been licensed one copy of this document for personal use only. Any other reproduction or redistribution is strictly prohibited. All rights reserved.
This content is provided for test preparation purposes only and does not imply an endorsement by Mometrix of any particular political, scientific, or religious point of view.

DAY 31

IDENTIFYING TIME

What time is it? Write the answer under each clock.

9:00

3:30

Copyright © Mometrix Media. You have been licensed one copy of this document for personal use only. Any other reproduction or redistribution is strictly prohibited. All rights reserved.
This content is provided for test preparation purposes only and does not imply an endorsement by Mometrix of any particular political, scientific, or religious point of view.

Day 32

Writing Time

Write the correct time in the clock on the right.

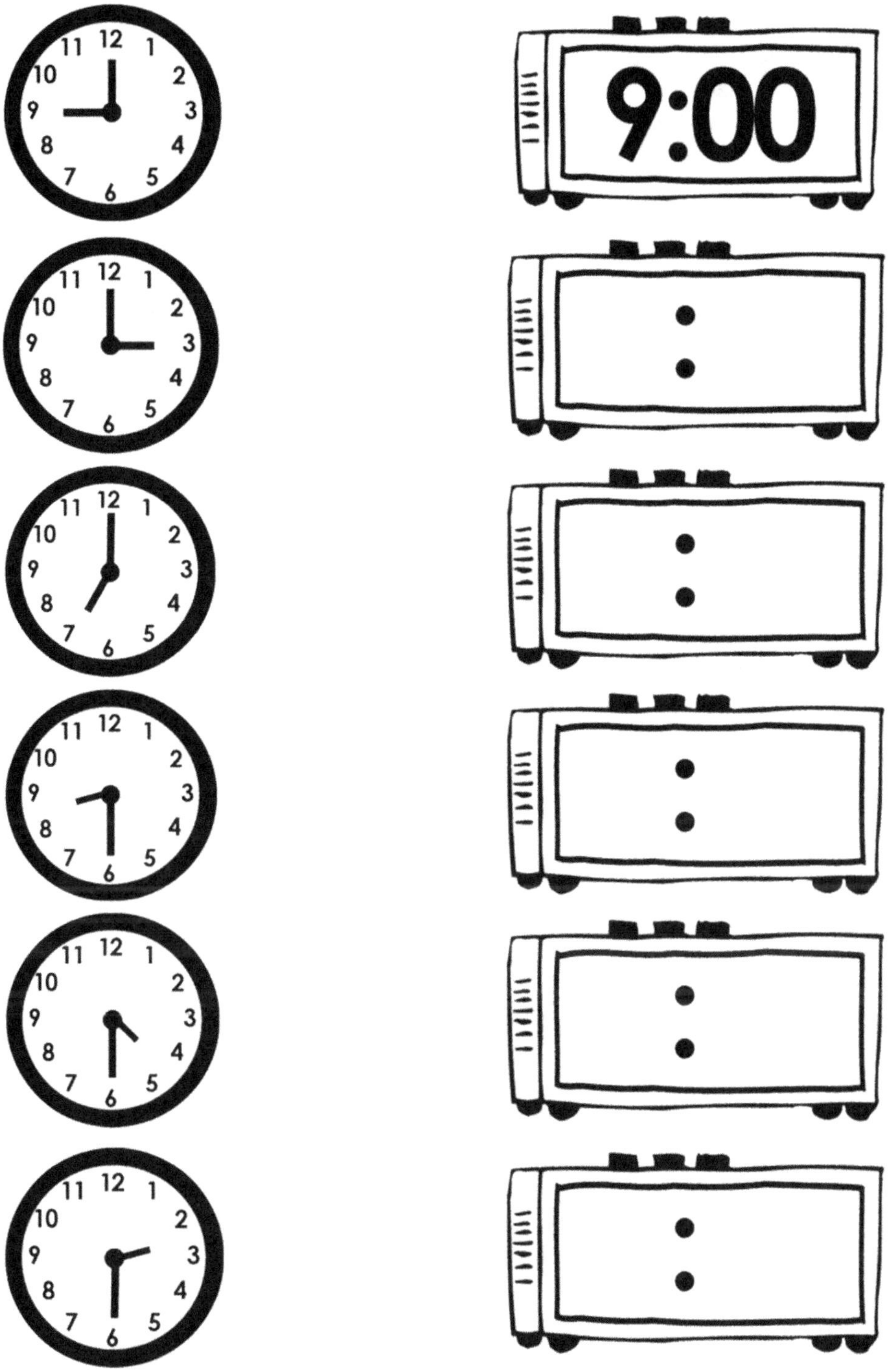

Copyright © Mometrix Media. You have been licensed one copy of this document for personal use only. Any other reproduction or redistribution is strictly prohibited. All rights reserved.
This content is provided for test preparation purposes only and does not imply an endorsement by Mometrix of any particular political, scientific, or religious point of view.

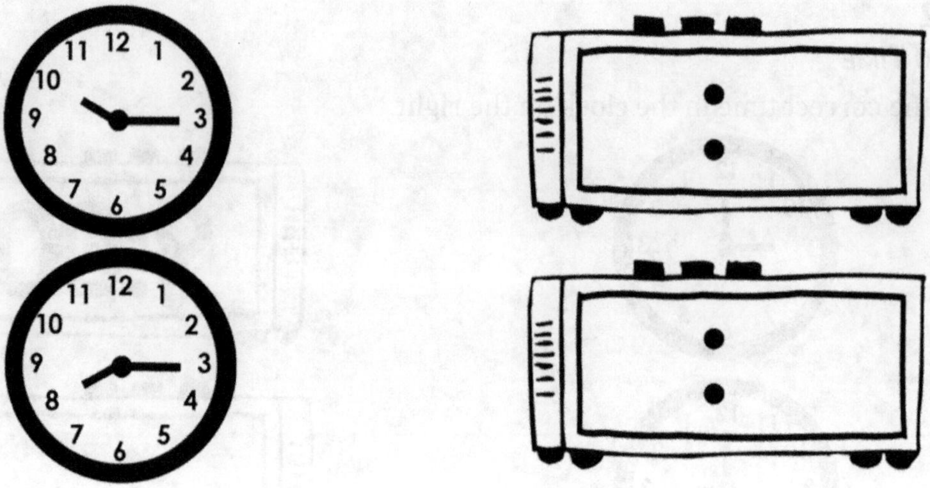

Copyright © Mometrix Media. You have been licensed one copy of this document for personal use only. Any other reproduction or redistribution is strictly prohibited. All rights reserved.
This content is provided for test preparation purposes only and does not imply an endorsement by Mometrix of any particular political, scientific, or religious point of view.

DAY 33

DRAWING TIME

Draw the correct time in the clock on the right.

Copyright © Mometrix Media. You have been licensed one copy of this document for personal use only. Any other reproduction or redistribution is strictly prohibited. All rights reserved.
This content is provided for test preparation purposes only and does not imply an endorsement by Mometrix of any particular political, scientific, or religious point of view.

Copyright © Mometrix Media. You have been licensed one copy of this document for personal use only. Any other reproduction or redistribution is strictly prohibited. All rights reserved.
This content is provided for test preparation purposes only and does not imply an endorsement by Mometrix of any particular political, scientific, or religious point of view.

Day 34

Time-Matching

Draw a line from each clock on the left to match the clock on the right.

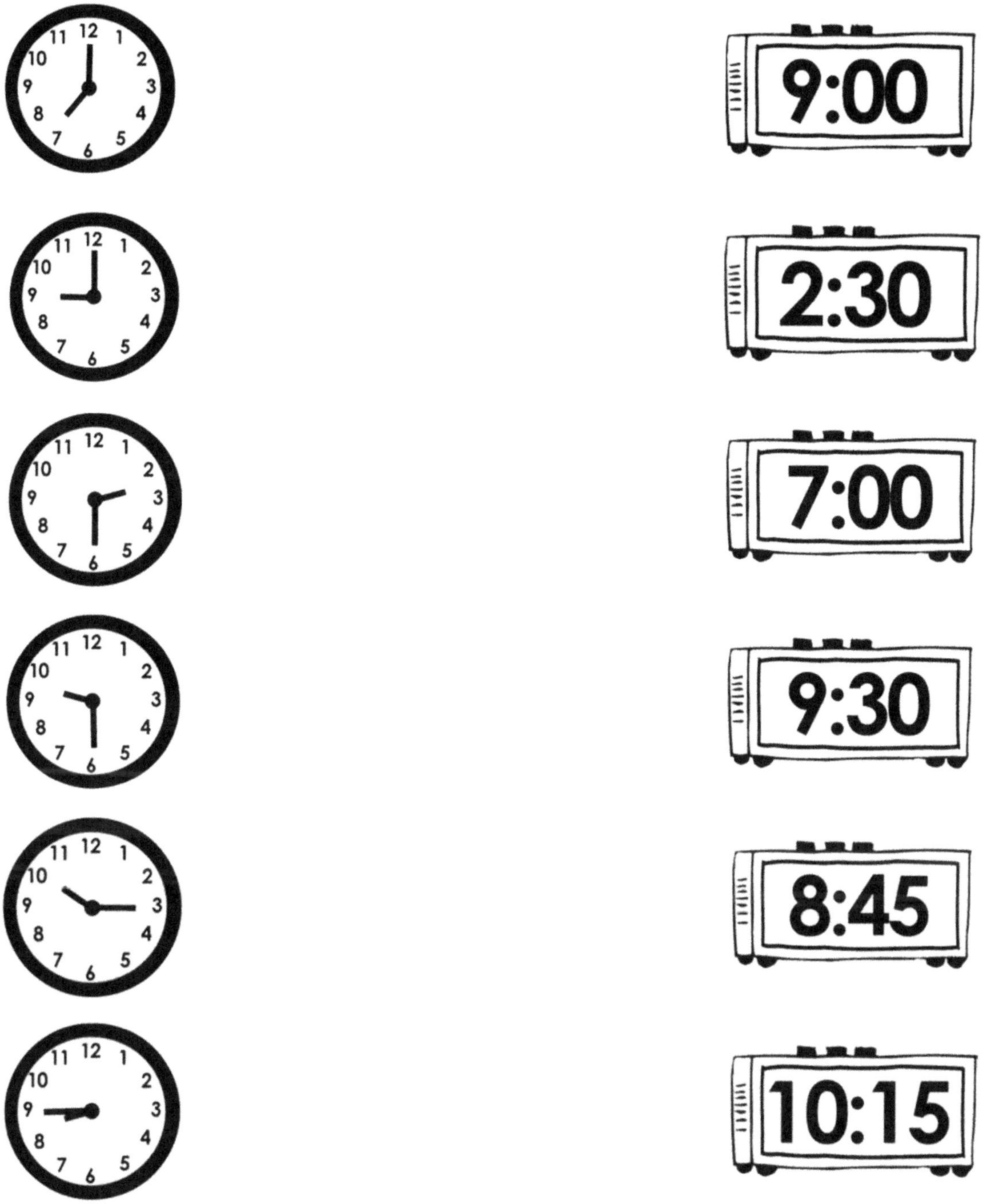

Copyright © Mometrix Media. You have been licensed one copy of this document for personal use only. Any other reproduction or redistribution is strictly prohibited. All rights reserved.
This content is provided for test preparation purposes only and does not imply an endorsement by Mometrix of any particular political, scientific, or religious point of view.

Chapter 6: Measurements

Measurements are used to determine how big, tall, long, heavy, or hot something is by comparing it to something else. Sometimes, objects are used to measure. Other times, precise measuring tools like a ruler or a scale are used. In the United States, the customary system of measurement is primarily used. Some basic units of measurement in this system include inches, feet, yards, miles, ounces, pounds, cups, and gallons.

Here are a few measuring tools that can be used.

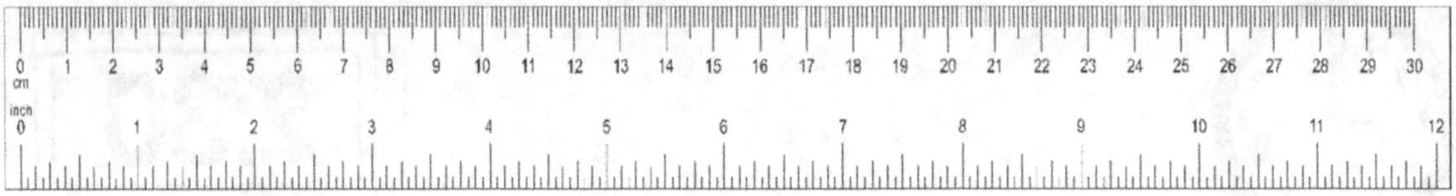

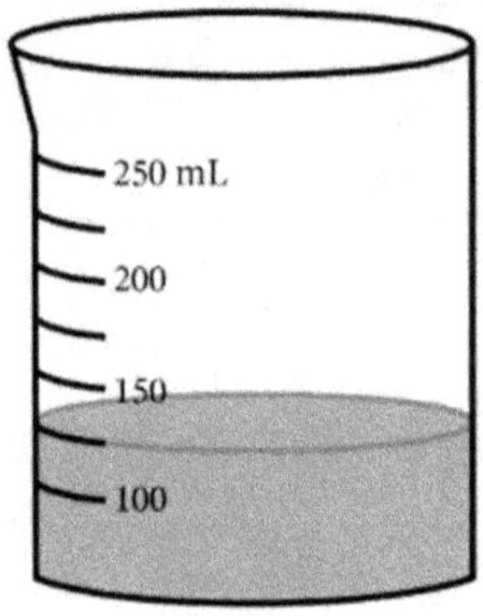

Complete one worksheet each day in this chapter. Each day should take no longer than 15 minutes to complete.

Copyright © Mometrix Media. You have been licensed one copy of this document for personal use only. Any other reproduction or redistribution is strictly prohibited. All rights reserved.
This content is provided for test preparation purposes only and does not imply an endorsement by Mometrix of any particular political, scientific, or religious point of view.

Day 35

Identifying Larger Objects

Which one is larger? Color the larger shape.

Copyright © Mometrix Media. You have been licensed one copy of this document for personal use only. Any other reproduction or redistribution is strictly prohibited. All rights reserved.
This content is provided for test preparation purposes only and does not imply an endorsement by Mometrix of any particular political, scientific, or religious point of view.

Day 36

Identifying Smaller Objects

Which one is smaller? Color the smaller shape.

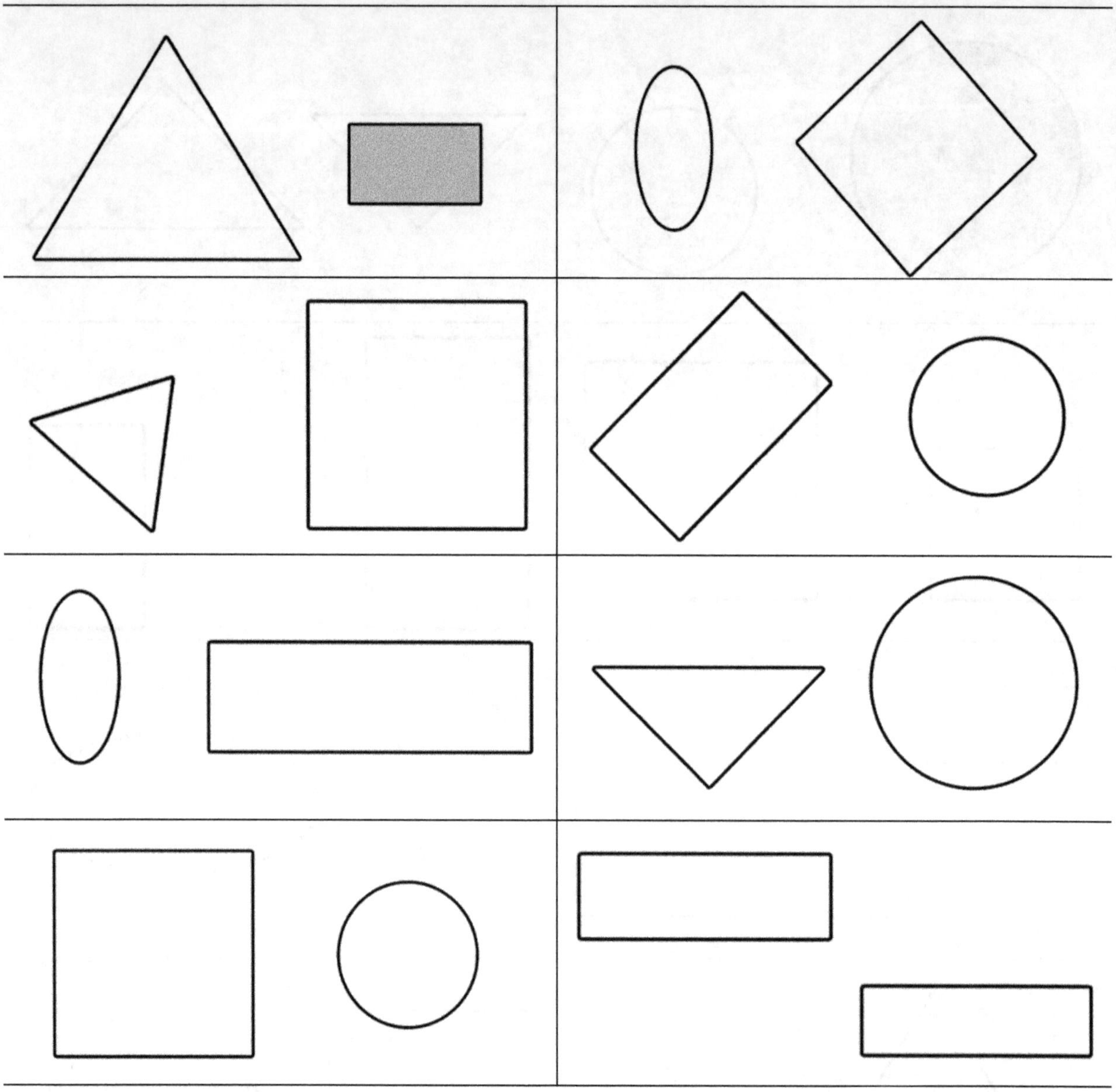

Copyright © Mometrix Media. You have been licensed one copy of this document for personal use only. Any other reproduction or redistribution is strictly prohibited. All rights reserved.
This content is provided for test preparation purposes only and does not imply an endorsement by Mometrix of any particular political, scientific, or religious point of view.

Day 37

Measuring Units: Length

= 1 Unit

Write the length of the objects in units.

5

Units

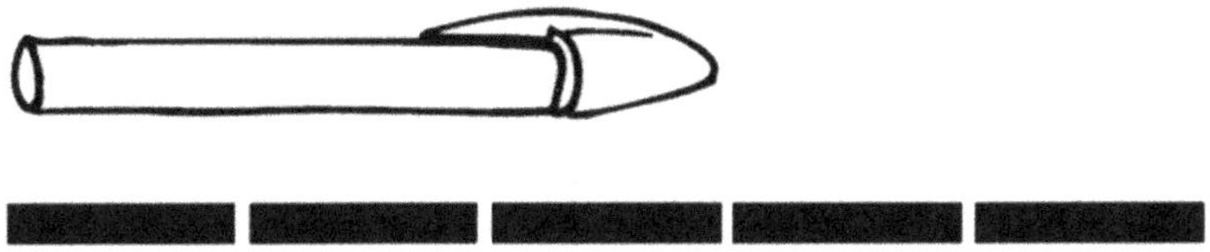

Units

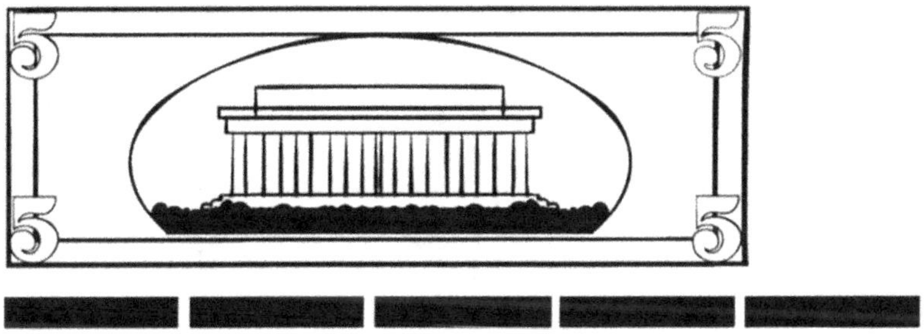

Units

Units

Copyright © Mometrix Media. You have been licensed one copy of this document for personal use only. Any other reproduction or redistribution is strictly prohibited. All rights reserved.
This content is provided for test preparation purposes only and does not imply an endorsement by Mometrix of any particular political, scientific, or religious point of view.

Day 38
Measuring Units: Height

= 1 Unit

Write the height of the objects in units.

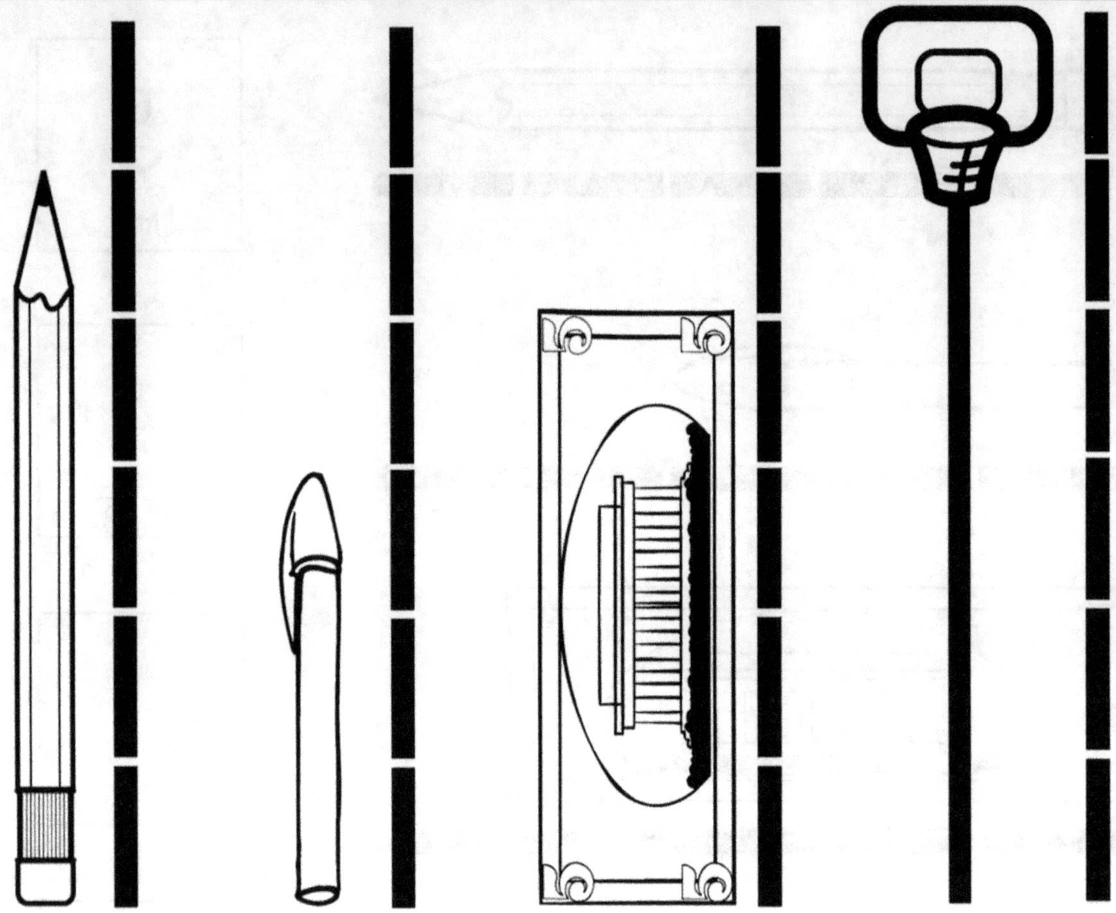

5 Units	Units	Units	Units

Copyright © Mometrix Media. You have been licensed one copy of this document for personal use only. Any other reproduction or redistribution is strictly prohibited. All rights reserved.
This content is provided for test preparation purposes only and does not imply an endorsement by Mometrix of any particular political, scientific, or religious point of view.

Day 39

Measuring Inches

Write the length of each object in the blanks.

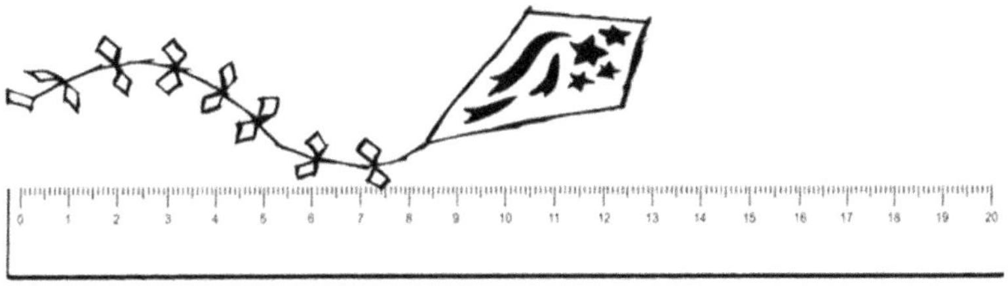

13 inches

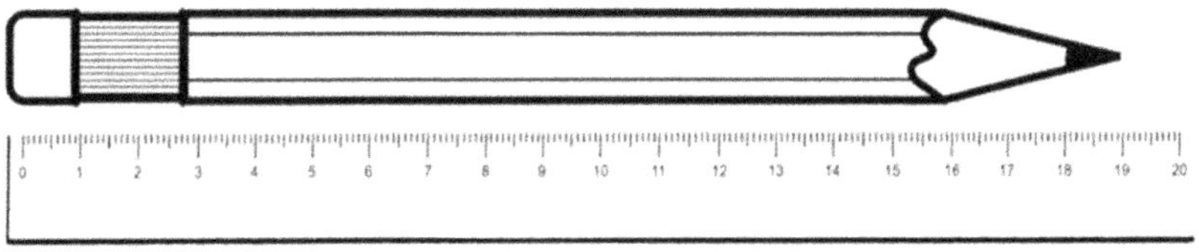

_____ inches

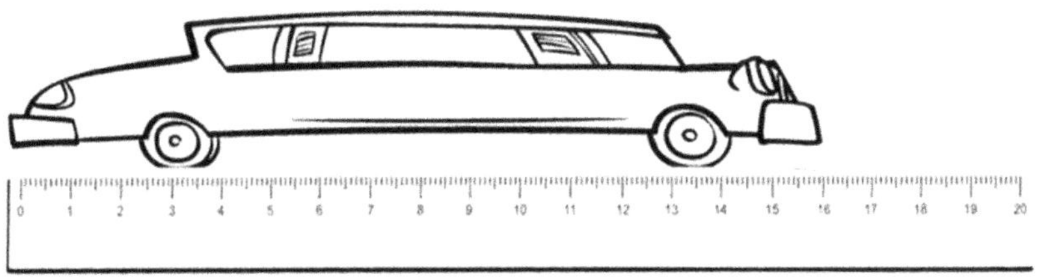

_____ inches

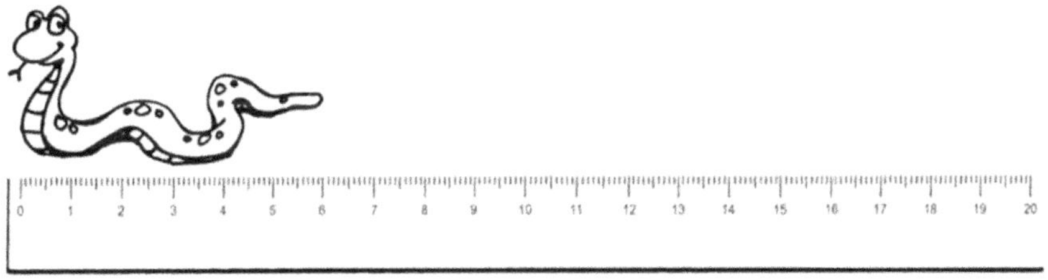

_____ inches

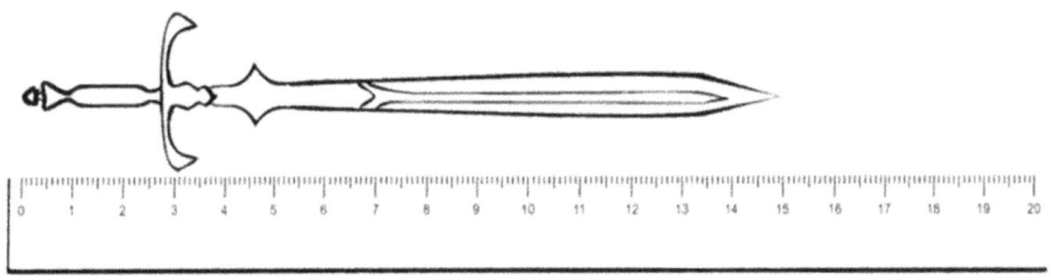

_____ inches

Copyright © Mometrix Media. You have been licensed one copy of this document for personal use only. Any other reproduction or redistribution is strictly prohibited. All rights reserved.
This content is provided for test preparation purposes only and does not imply an endorsement by Mometrix of any particular political, scientific, or religious point of view.

Day 40

Measuring Half-Inches

Write the length of each object in the blanks.

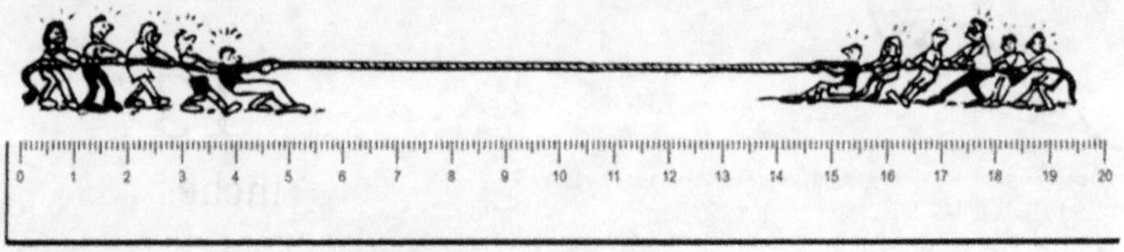

$19\frac{1}{2}$ inches

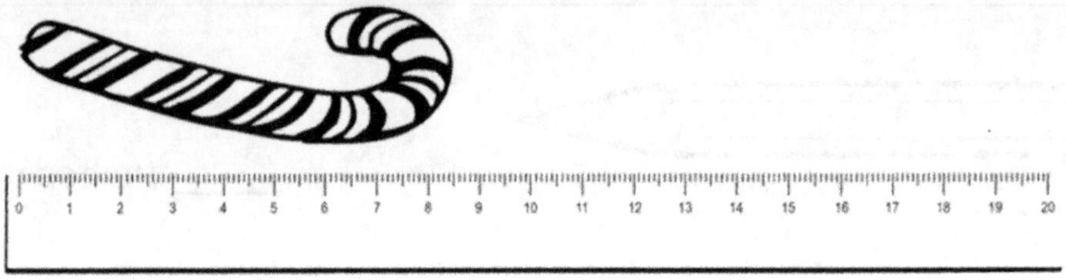

_____ inches

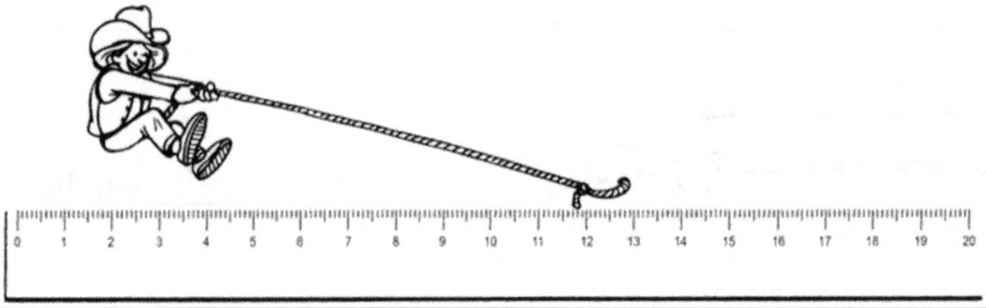

_____ inches

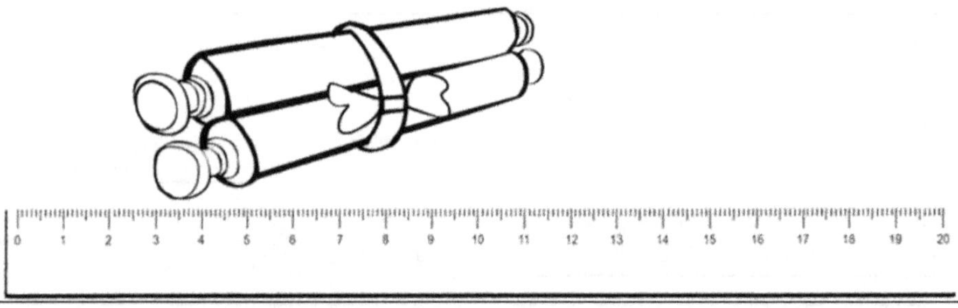

_____ inches

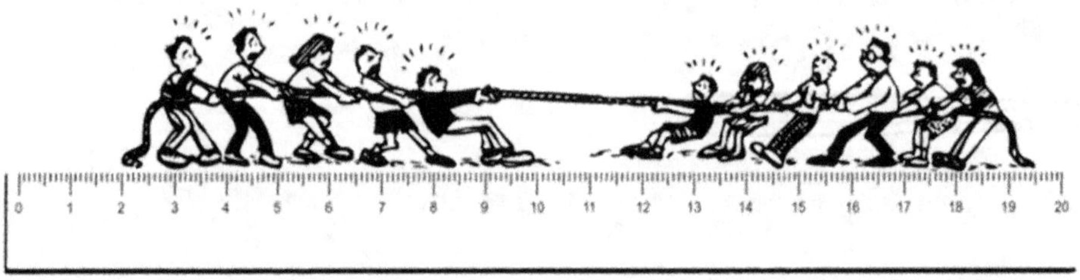

_____ inches

Copyright © Mometrix Media. You have been licensed one copy of this document for personal use only. Any other reproduction or redistribution is strictly prohibited. All rights reserved.
This content is provided for test preparation purposes only and does not imply an endorsement by Mometrix of any particular political, scientific, or religious point of view.

Day 41

Measuring Height

Write the height of each object in the box.

14 feet	feet	feet

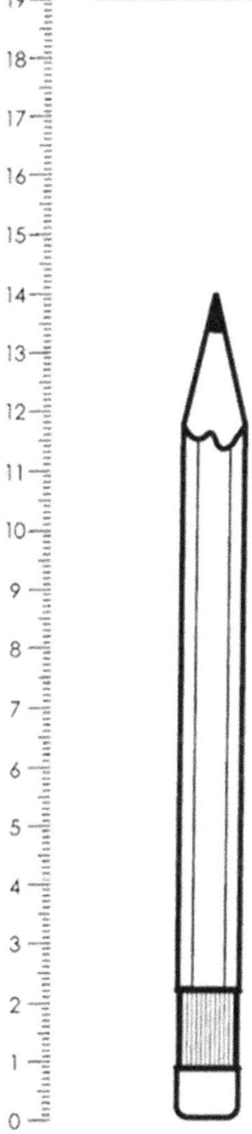

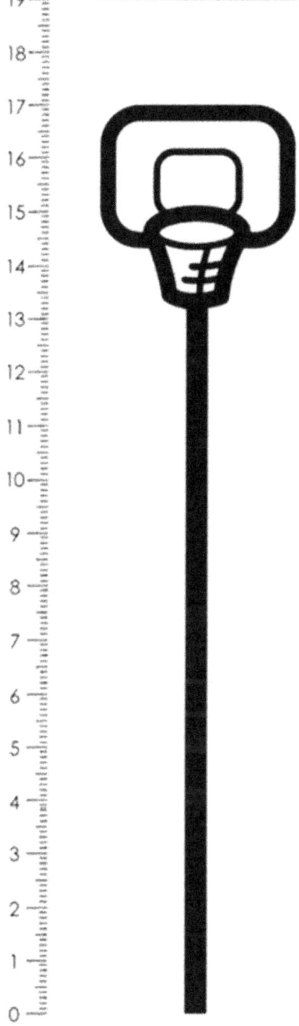

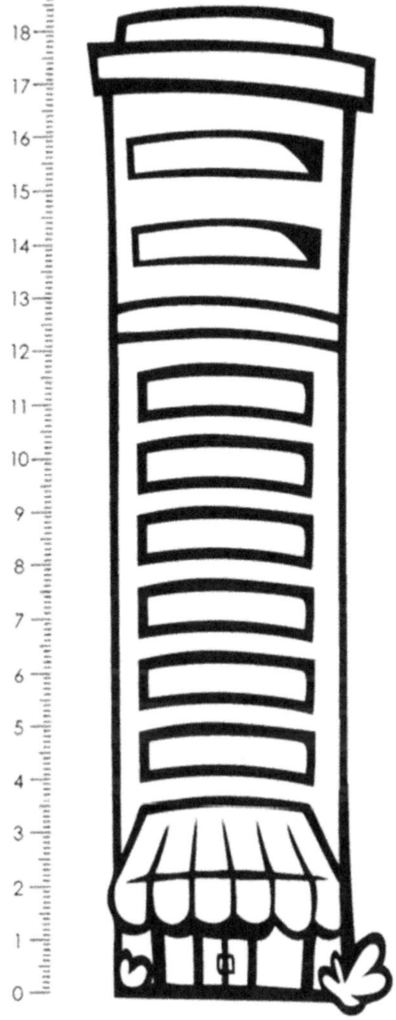

Copyright © Mometrix Media. You have been licensed one copy of this document for personal use only. Any other reproduction or redistribution is strictly prohibited. All rights reserved.
This content is provided for test preparation purposes only and does not imply an endorsement by Mometrix of any particular political, scientific, or religious point of view.

Day 42

Identifying Object Weights

Circle the object that weighs more.

Copyright © Mometrix Media. You have been licensed one copy of this document for personal use only. Any other reproduction or redistribution is strictly prohibited. All rights reserved.
This content is provided for test preparation purposes only and does not imply an endorsement by Mometrix of any particular political, scientific, or religious point of view.

DAY 43

WRITING OBJECT WEIGHTS

Write the correct weight in the box under each scale.

3
Pounds

Pounds

Pounds

Pounds

Pounds

Pounds

Copyright © Mometrix Media. You have been licensed one copy of this document for personal use only. Any other reproduction or redistribution is strictly prohibited. All rights reserved.
This content is provided for test preparation purposes only and does not imply an endorsement by Mometrix of any particular political, scientific, or religious point of view.

Day 44

Writing Temperature

Write the correct temperature of each thermometer in the blanks.

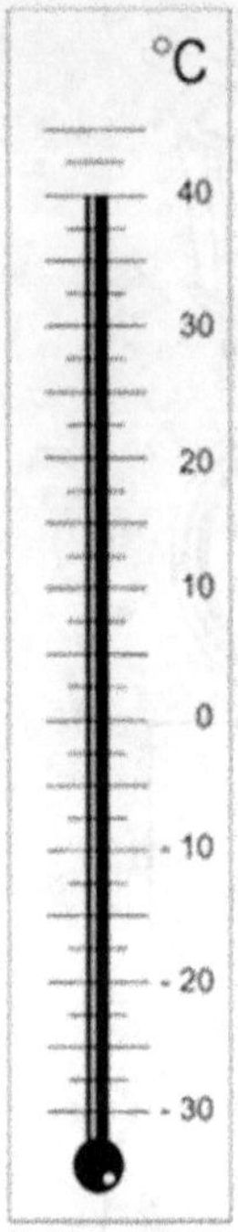

40°

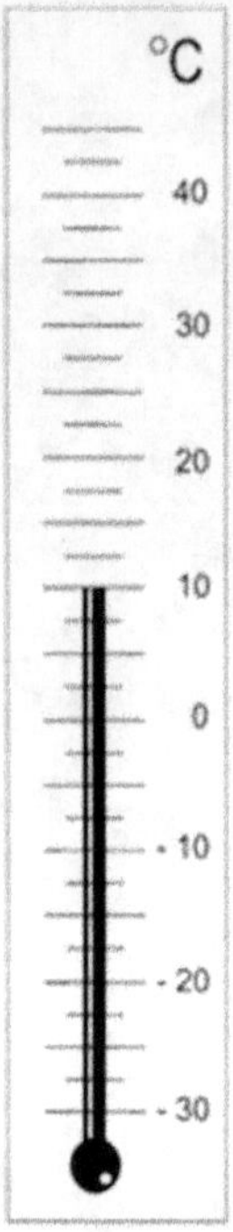

______

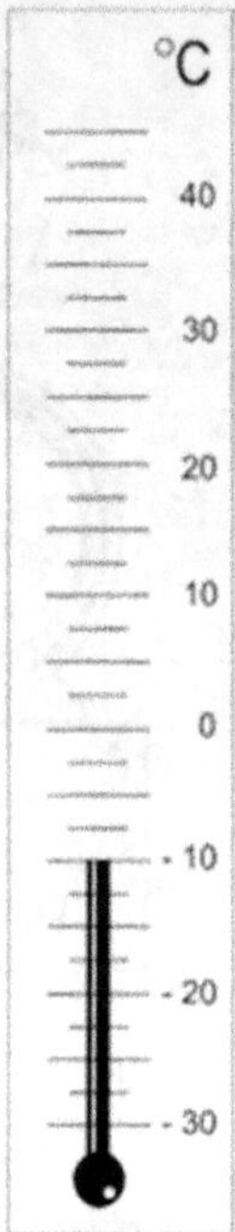

______

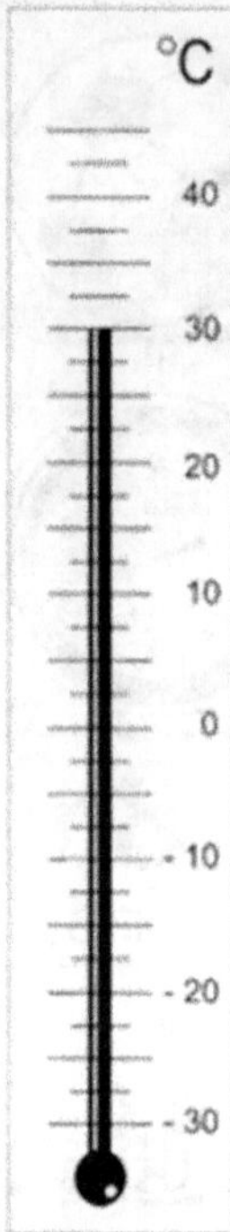

______

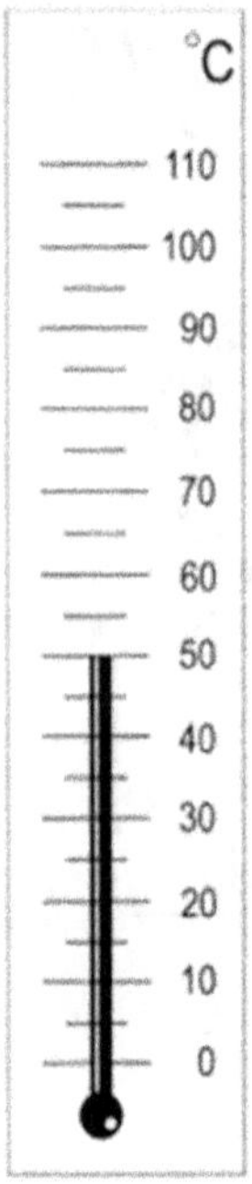

______

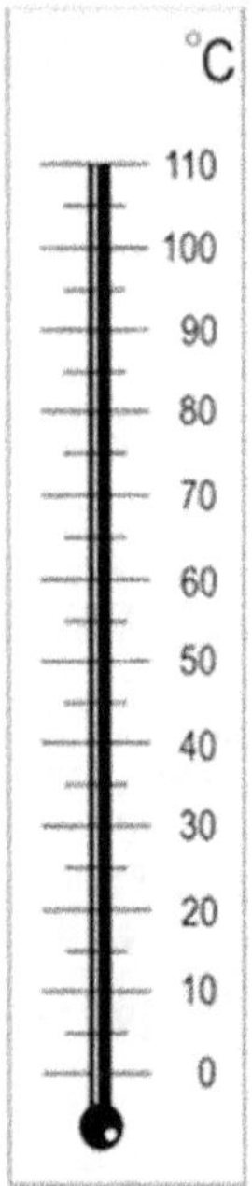

______

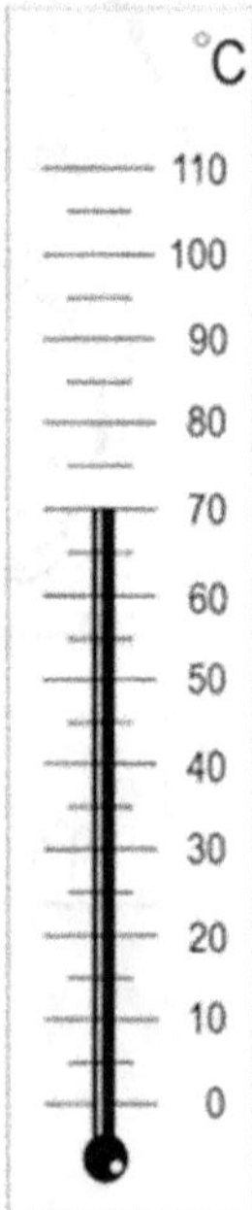

______

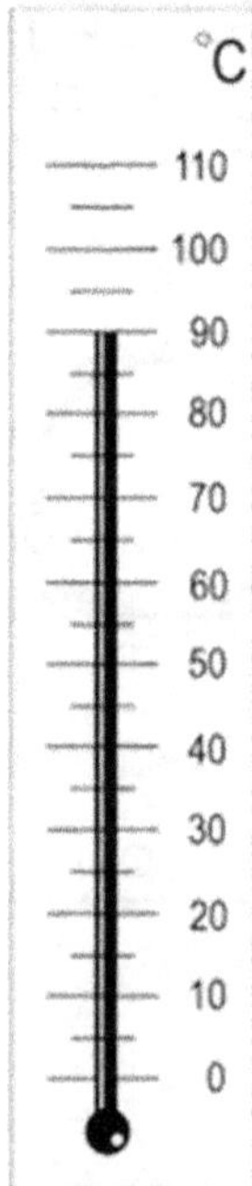

______

Copyright © Mometrix Media. You have been licensed one copy of this document for personal use only. Any other reproduction or redistribution is strictly prohibited. All rights reserved.
This content is provided for test preparation purposes only and does not imply an endorsement by Mometrix of any particular political, scientific, or religious point of view.

Chapter 7: Geometry

Two-dimensional shapes can be classified by the number of sides they have. A polygon is considered regular if all the sides are the same length. For example, a regular pentagon has 5 sides, and all of the sides are the same length. Side lengths that are the same are considered **congruent**. For example, a square has 4 congruent side lengths.

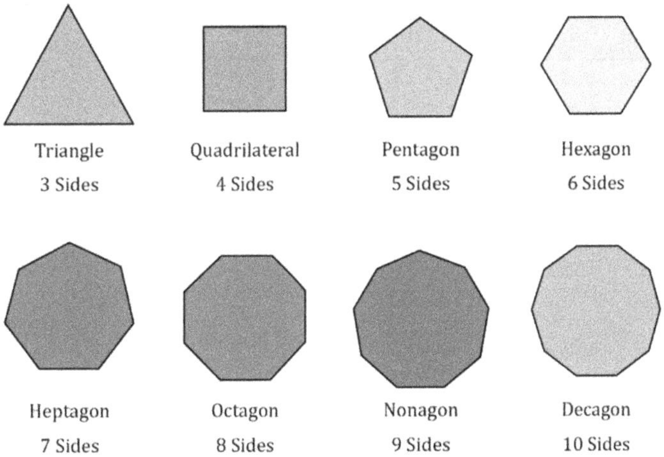

The most common three-dimensional figures include cones, cylinders, spheres, triangular prims, rectangular prisms, and cubes. When identifying and classifying three-dimensional figures, it is important to be aware of the number of faces, edges, and vertices. **Faces** are the flat surfaces, **edges** are where two faces come together, and **vertices** are corners where more than two faces come together. For example, a rectangular prism has 6 faces, 12 edges, and 8 vertices.

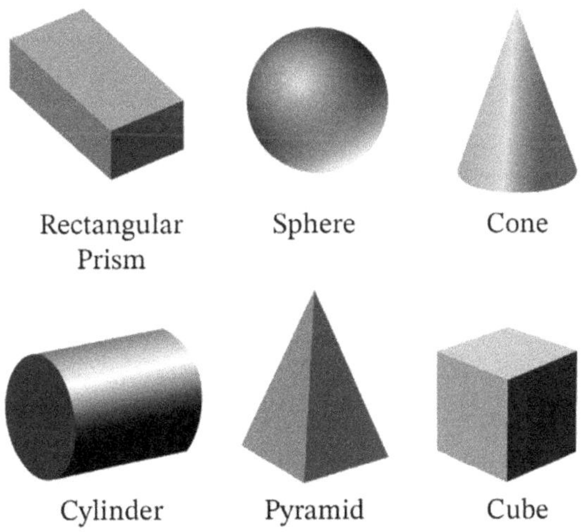

Complete one worksheet each day in this chapter. Each day should take no longer than 15 minutes to complete.

Copyright © Mometrix Media. You have been licensed one copy of this document for personal use only. Any other reproduction or redistribution is strictly prohibited. All rights reserved.
This content is provided for test preparation purposes only and does not imply an endorsement by Mometrix of any particular political, scientific, or religious point of view.

Day 45
Writing Shapes

	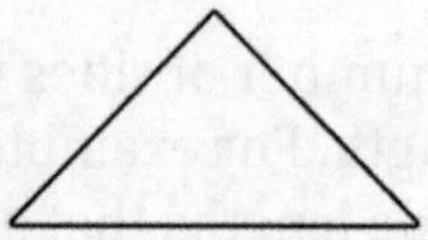	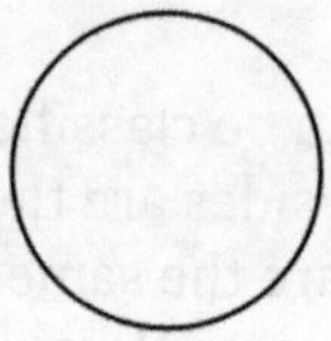	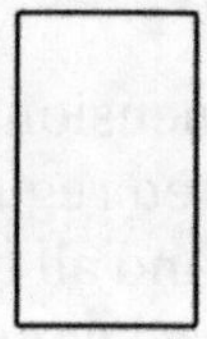
Square	Triangle	Circle	Rectangle

Write the names of the shapes in the blanks.

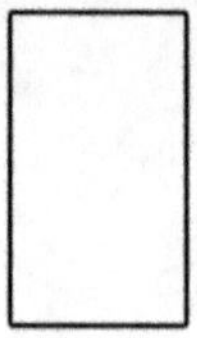	Rectangle	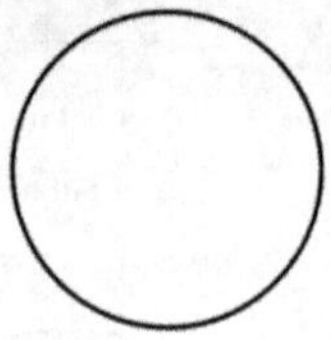	________
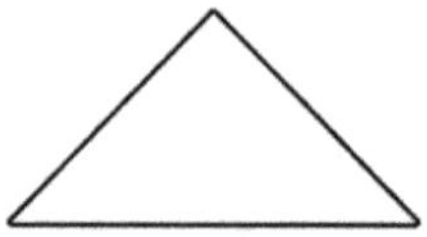	________		________
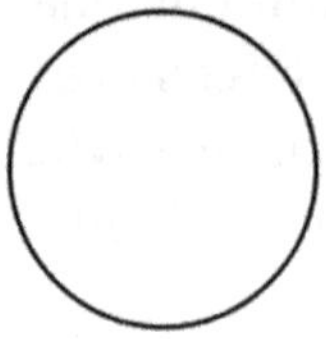	________	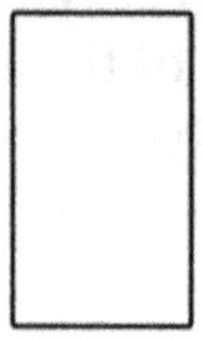	________
	________	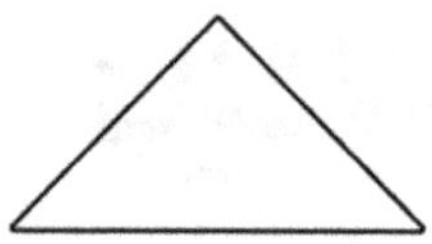	________

Copyright © Mometrix Media. You have been licensed one copy of this document for personal use only. Any other reproduction or redistribution is strictly prohibited. All rights reserved.
This content is provided for test preparation purposes only and does not imply an endorsement by Mometrix of any particular political, scientific, or religious point of view.

Day 46
Drawing Shapes

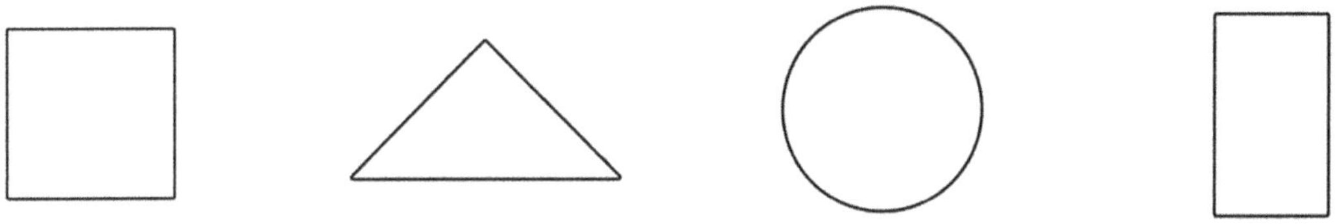

Square   Triangle   Circle   Rectangle

Draw each shape below.

Square 	Triangle
Circle	Rectangle
Triangle	Circle
Square	Rectangle

Copyright © Mometrix Media. You have been licensed one copy of this document for personal use only. Any other reproduction or redistribution is strictly prohibited. All rights reserved.
This content is provided for test preparation purposes only and does not imply an endorsement by Mometrix of any particular political, scientific, or religious point of view.

Day 47

Counting Sides of Shapes

How many sides does each shape have?

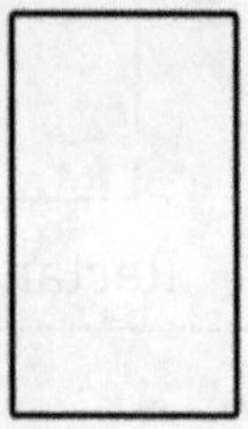	4	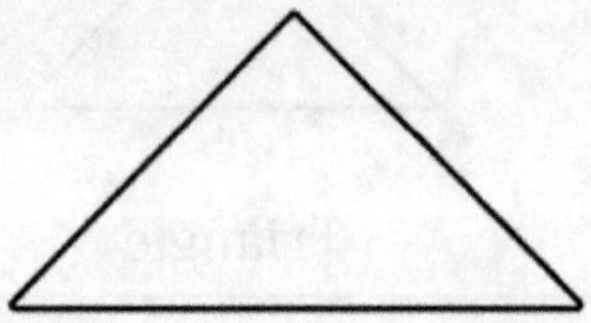	____
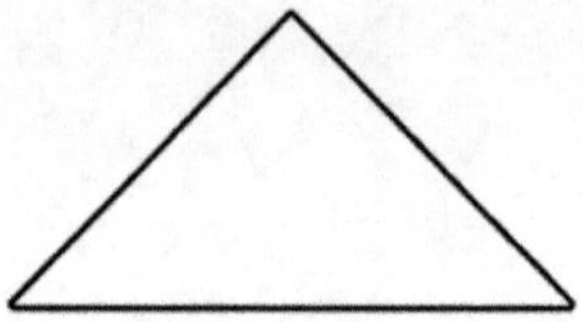	____	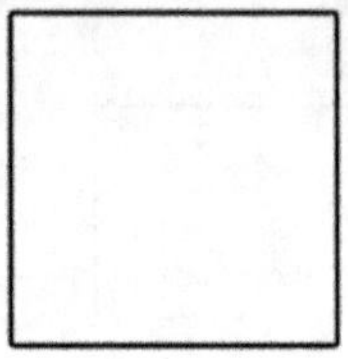	____
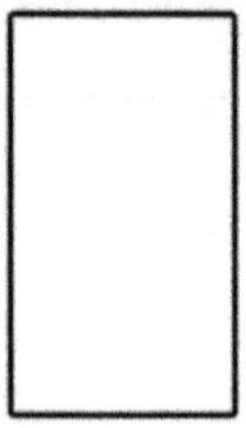	____	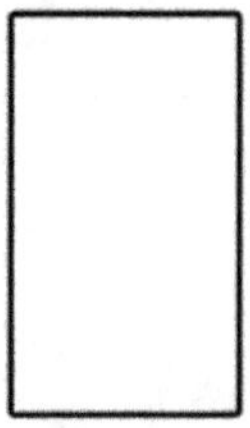	____
	____	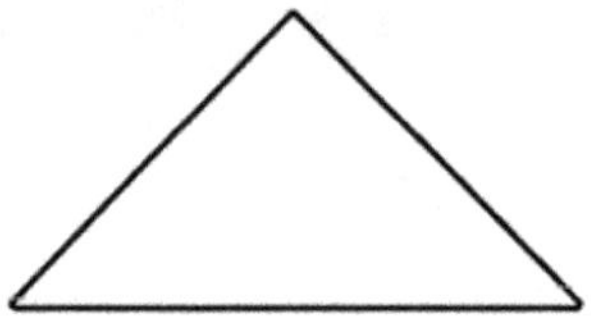	____
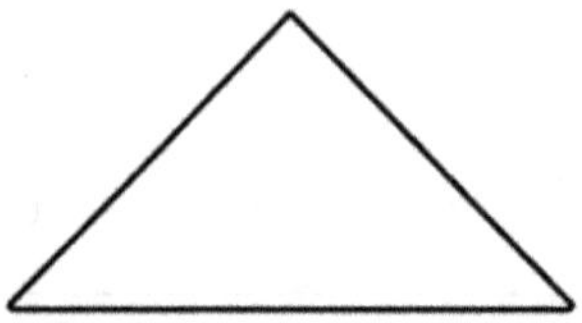	____	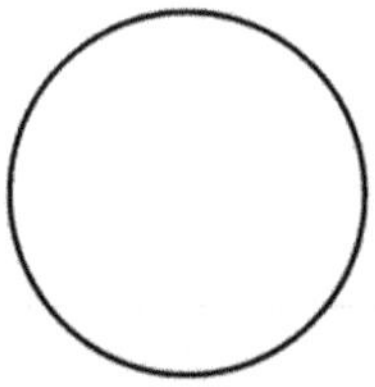	____

Copyright © Mometrix Media. You have been licensed one copy of this document for personal use only. Any other reproduction or redistribution is strictly prohibited. All rights reserved.
This content is provided for test preparation purposes only and does not imply an endorsement by Mometrix of any particular political, scientific, or religious point of view.

Day 48

Writing Advanced Shapes

Sphere

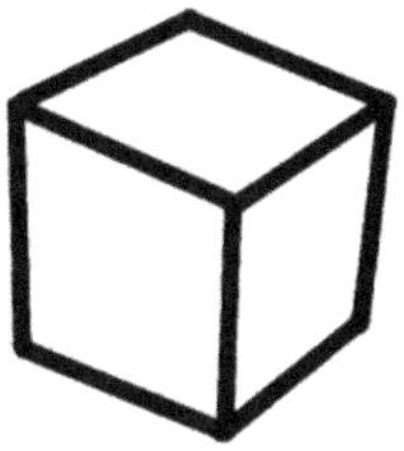
Cube

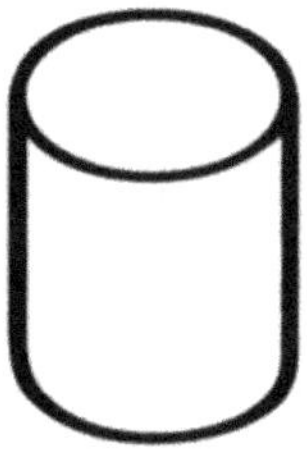
Cylinder

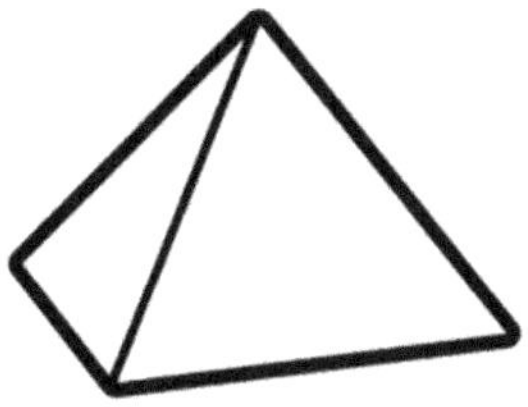
Pyramid

Write the names of the shapes in the blanks.

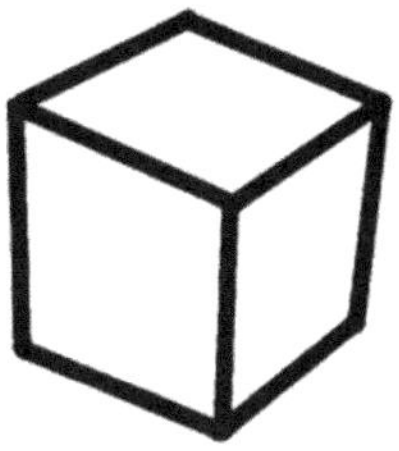 Cube

 ________

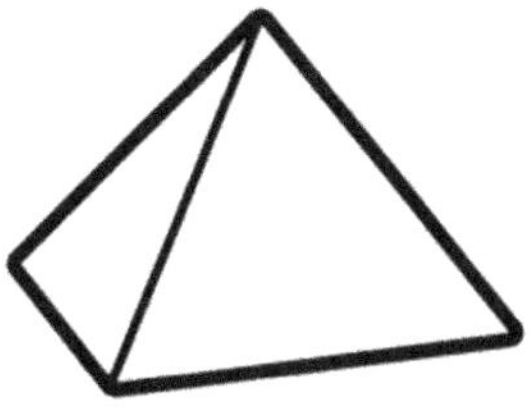 ________

 ________

 ________

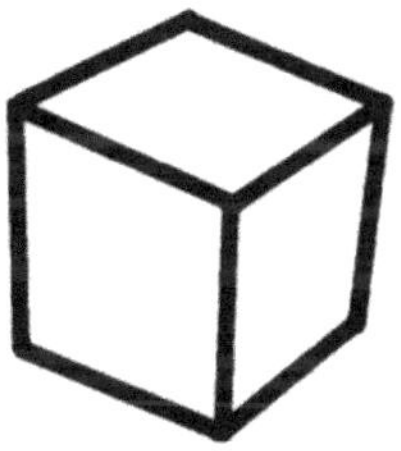 ________

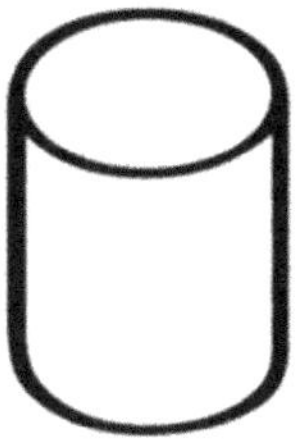 ________

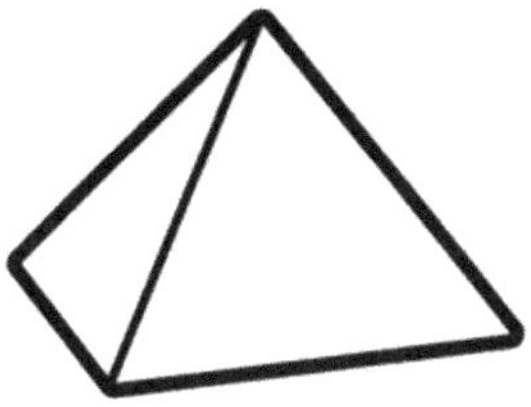 ________

Copyright © Mometrix Media. You have been licensed one copy of this document for personal use only. Any other reproduction or redistribution is strictly prohibited. All rights reserved.
This content is provided for test preparation purposes only and does not imply an endorsement by Mometrix of any particular political, scientific, or religious point of view.

Day 49

Counting & Coloring Shapes

Color all the rectangles red. How many rectangles did you count? ________

Color all the circles blue. How many circles did you count? ________

Color all the squares green. How many squares did you count? ________

Color all the triangles orange. How many triangles did you count? ________

Copyright © Mometrix Media. You have been licensed one copy of this document for personal use only. Any other reproduction or redistribution is strictly prohibited. All rights reserved.
This content is provided for test preparation purposes only and does not imply an endorsement by Mometrix of any particular political, scientific, or religious point of view.

DAY 50

COUNTING & COLORING ADVANCED SHAPES

Color all the spheres red. How many spheres did you count? ________

Color all the cubes blue. How many cubes did you count? ________

Color all the cylinders green. How many cylinders did you count? ________

Color all the pyramids orange. How many pyramids did you count? ________

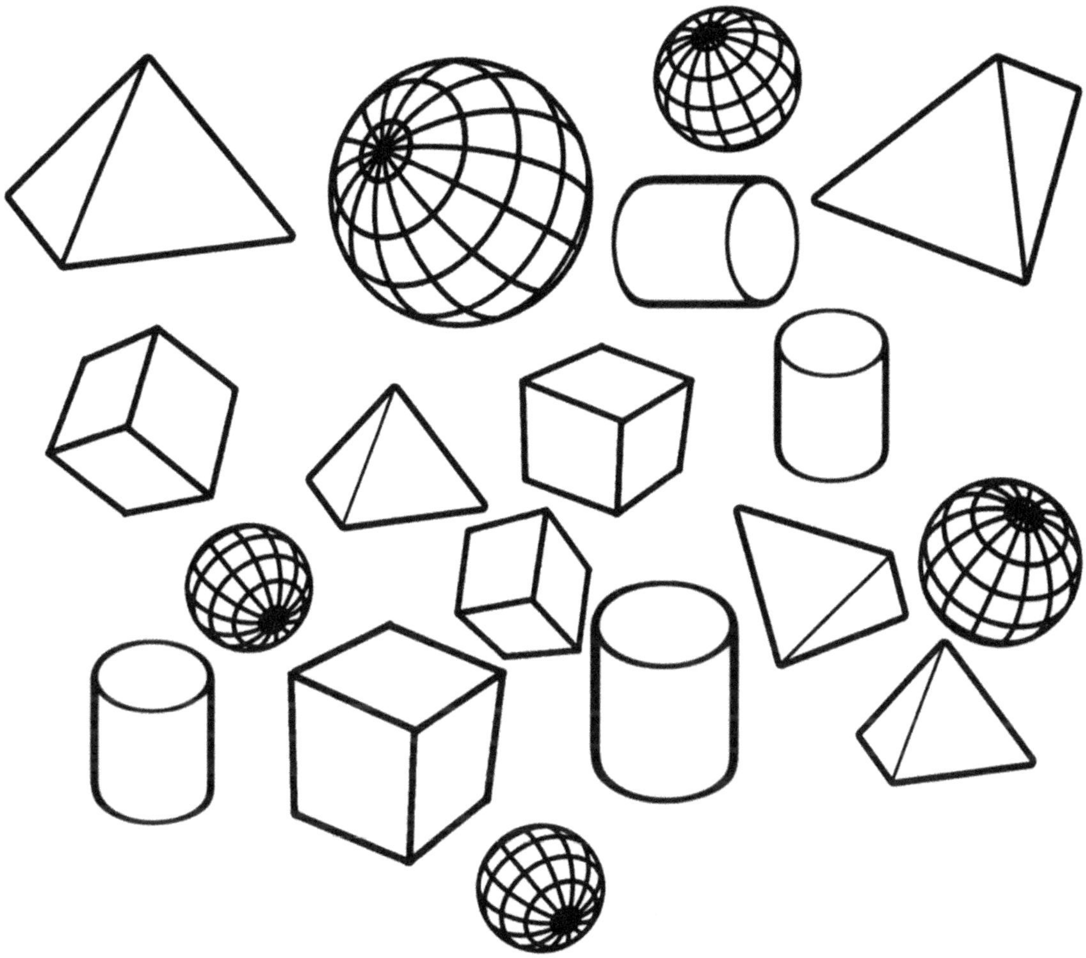

Copyright © Mometrix Media. You have been licensed one copy of this document for personal use only. Any other reproduction or redistribution is strictly prohibited. All rights reserved.
This content is provided for test preparation purposes only and does not imply an endorsement by Mometrix of any particular political, scientific, or religious point of view.

Workbook Answers

Chapter 1: Counting

Day 1

Counting to 10

1. 10

2. 9

3. 10

4. 8

5. 9

6. 8

Day 2

Color the Objects to 10

1. 

2. 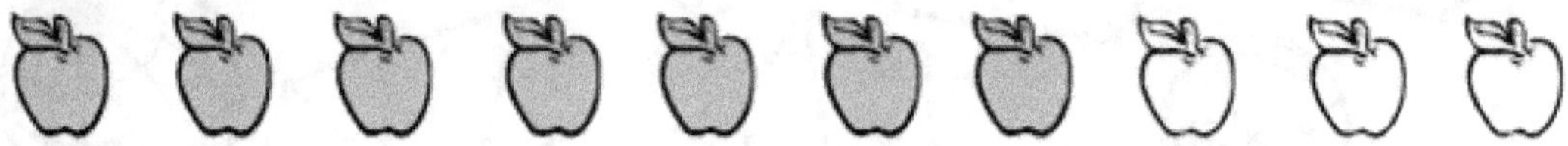

3. 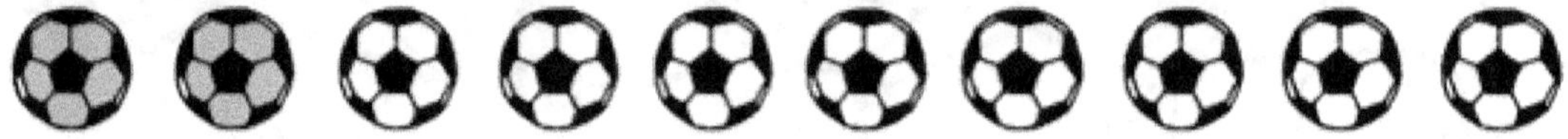

4. 

5. 

6. 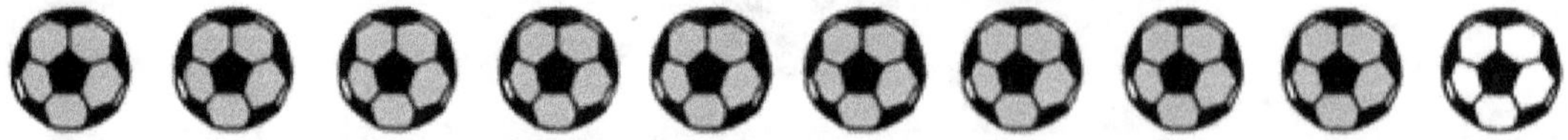

7. 

8. 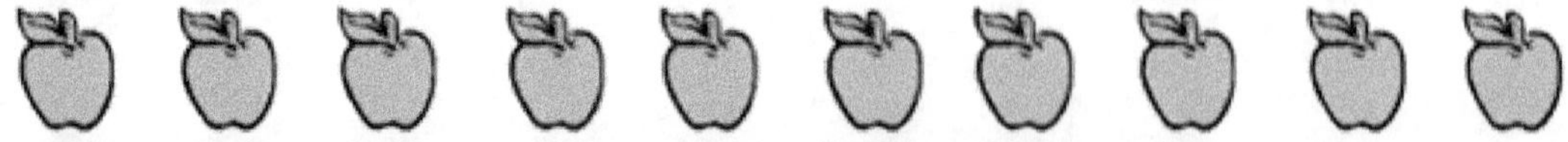

9. 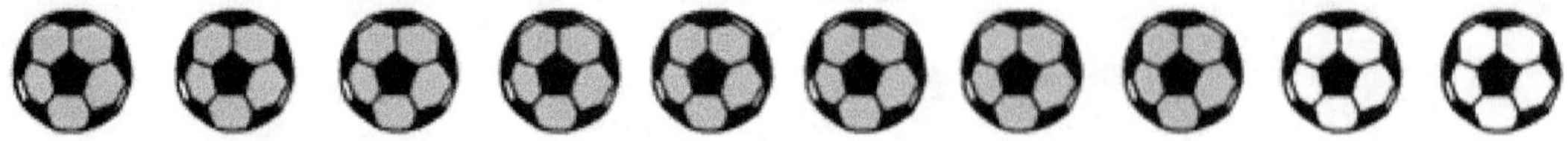

10. 

Copyright © Mometrix Media. You have been licensed one copy of this document for personal use only. Any other reproduction or redistribution is strictly prohibited. All rights reserved.
This content is provided for test preparation purposes only and does not imply an endorsement by Mometrix of any particular political, scientific, or religious point of view.

11. 

Day 3

Counting & Writing Numbers 1-10 Part 1

1. 5

five

2. 10

ten

3. 7

seven

4. 8

eight

Counting & Writing Numbers 1-10 Part 2

1. 10

ten

2. 5

five

3. 4

four

Copyright © Mometrix Media. You have been licensed one copy of this document for personal use only. Any other reproduction or redistribution is strictly prohibited. All rights reserved.
This content is provided for test preparation purposes only and does not imply an endorsement by Mometrix of any particular political, scientific, or religious point of view.

4. 8

eight

Day 4
Counting by 2's

2	4	6	8	10

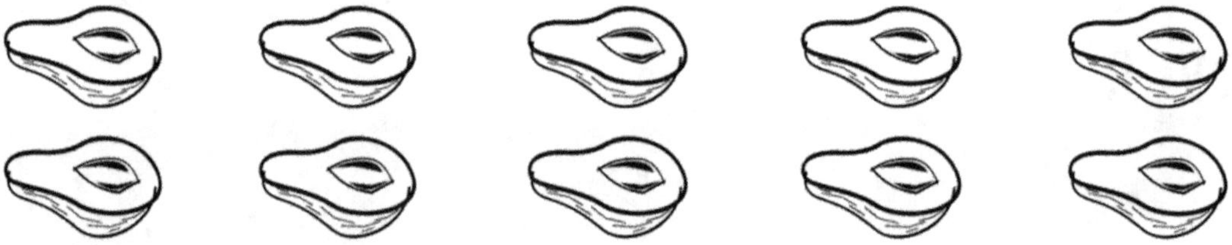

2	4	6	8	10

2	4	6	8	10

Copyright © Mometrix Media. You have been licensed one copy of this document for personal use only. Any other reproduction or redistribution is strictly prohibited. All rights reserved.
This content is provided for test preparation purposes only and does not imply an endorsement by Mometrix of any particular political, scientific, or religious point of view.

Day 5
Counting 1-10

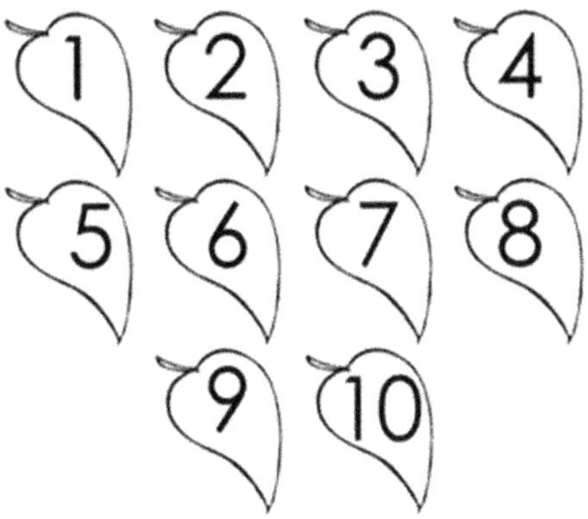

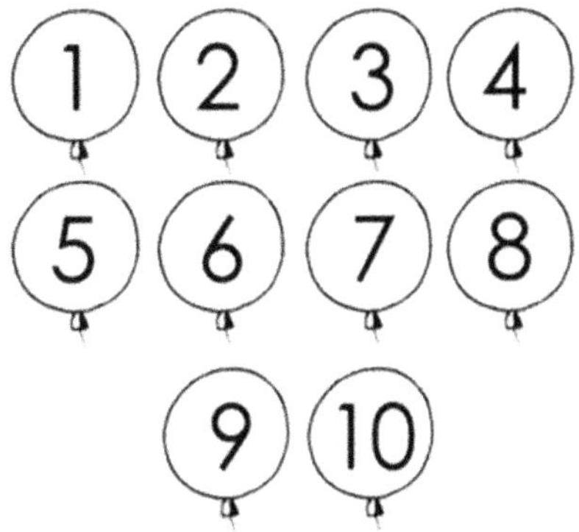

Copyright © Mometrix Media. You have been licensed one copy of this document for personal use only. Any other reproduction or redistribution is strictly prohibited. All rights reserved.
This content is provided for test preparation purposes only and does not imply an endorsement by Mometrix of any particular political, scientific, or religious point of view.

DAY 6

COUNTING OBJECTS 1-20

15

20

17

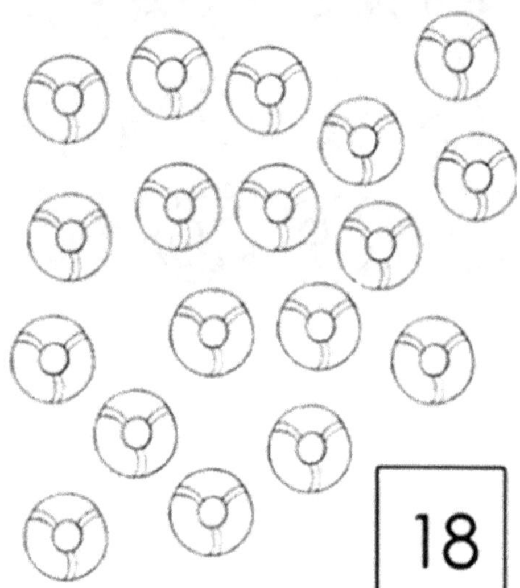

18

Copyright © Mometrix Media. You have been licensed one copy of this document for personal use only. Any other reproduction or redistribution is strictly prohibited. All rights reserved.
This content is provided for test preparation purposes only and does not imply an endorsement by Mometrix of any particular political, scientific, or religious point of view.

Day 7

Counting Objects 1-50

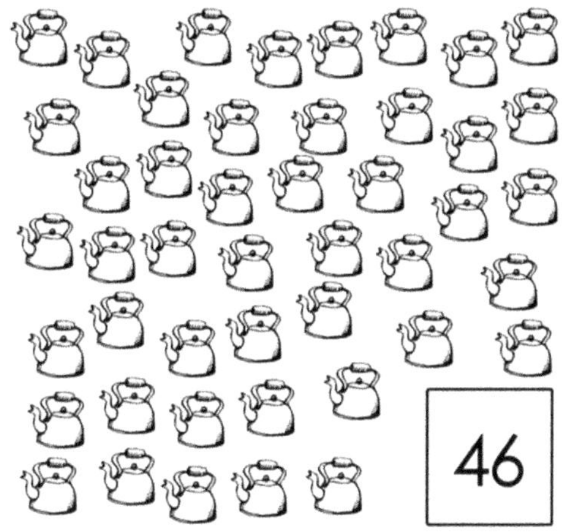

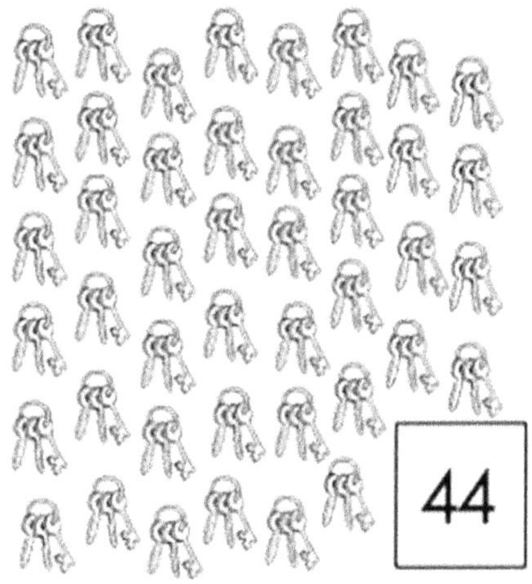

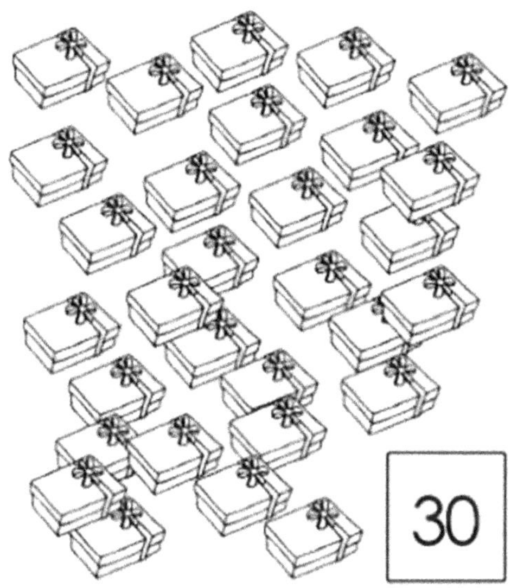

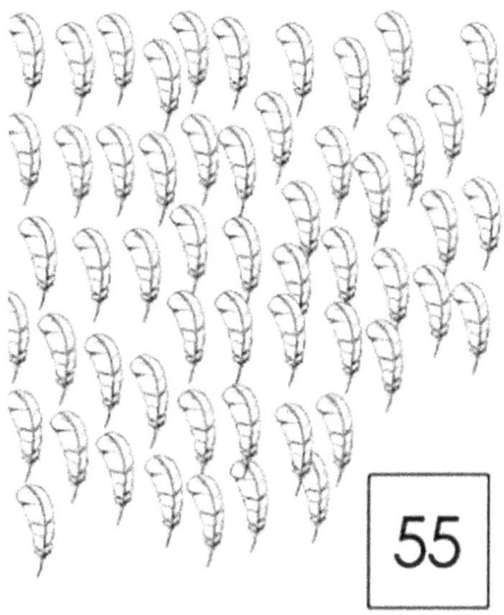

Copyright © Mometrix Media. You have been licensed one copy of this document for personal use only. Any other reproduction or redistribution is strictly prohibited. All rights reserved.
This content is provided for test preparation purposes only and does not imply an endorsement by Mometrix of any particular political, scientific, or religious point of view.

Day 8

Counting 1-120

1	2	3	4	5	6	7	8	9	10
11	12	13	14	15	16	17	18	19	20
21	22	23	24	25	26	27	28	29	30
31	32	33	34	35	36	37	38	39	40
41	42	43	44	45	46	47	48	49	50
51	52	53	54	55	56	57	58	59	60
61	62	63	64	65	66	67	68	69	70
71	72	73	74	75	76	77	78	79	80
81	82	83	84	85	86	87	88	89	90
91	92	93	94	95	96	97	98	99	100
101	102	103	104	105	106	107	108	109	110
111	112	113	114	115	116	117	118	119	120

Copyright © Mometrix Media. You have been licensed one copy of this document for personal use only. Any other reproduction or redistribution is strictly prohibited. All rights reserved.
This content is provided for test preparation purposes only and does not imply an endorsement by Mometrix of any particular political, scientific, or religious point of view.

Chapter 2: Addition

Day 9

Addition to 5

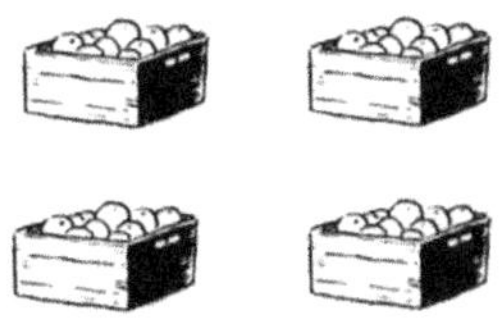 +  = 5

 + 0 = 3

 +  +  = 3

 +  +  +  = 4

 +  = 5

Copyright © Mometrix Media. You have been licensed one copy of this document for personal use only. Any other reproduction or redistribution is strictly prohibited. All rights reserved.
This content is provided for test preparation purposes only and does not imply an endorsement by Mometrix of any particular political, scientific, or religious point of view.

Day 10

Addition Groups 1-10

 +  $5 + 4 = 9$	 +  $2 + 2 = 4$
 + 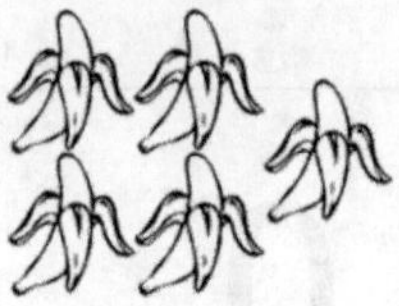 $3 + 5 = 8$	 +  $4 + 3 = 7$
 +  $6 + 2 = 8$	 + 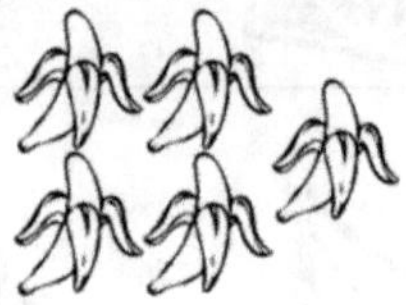 $1 + 5 = 6$
 +  $7 + 2 = 9$	 +  $8 + 2 = 10$

Copyright © Mometrix Media. You have been licensed one copy of this document for personal use only. Any other reproduction or redistribution is strictly prohibited. All rights reserved.
This content is provided for test preparation purposes only and does not imply an endorsement by Mometrix of any particular political, scientific, or religious point of view.

Day 11

Word Problems 1-10

We see 5 

1 more  shows up

5
+ 1

6

What is the sum? 6

There is 1 

9 more  arrive

1
+ 9

10

What is the total? 10

There are 4 

4 more  show up

4
+ 4

8

What is the sum? 8

We have 3

1 swims up

3
+ 1

4

How many are there? 4

There are 2 

7 more  arrive

2
+ 7

9

What is the total? 9

There are 3 

5 more  walk up

3
+ 5

8

What is the sum? 8

Jimmy has 2 

Sue has 4 

2
+ 4

6

What is the sum? 6

Steve has 5 

He added 2 

5
+ 2

7

What is the total? 7

Workbook Answers

Copyright © Mometrix Media. You have been licensed one copy of this document for personal use only. Any other reproduction or redistribution is strictly prohibited. All rights reserved.
This content is provided for test preparation purposes only and does not imply an endorsement by Mometrix of any particular political, scientific, or religious point of view.

Day 12

Addition 1-10

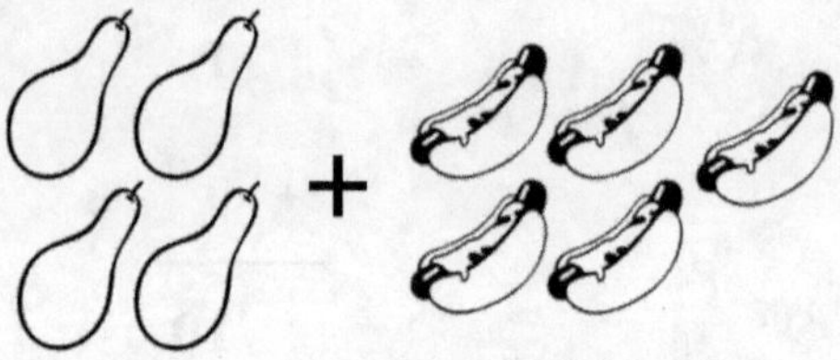

4 + 5 = 9

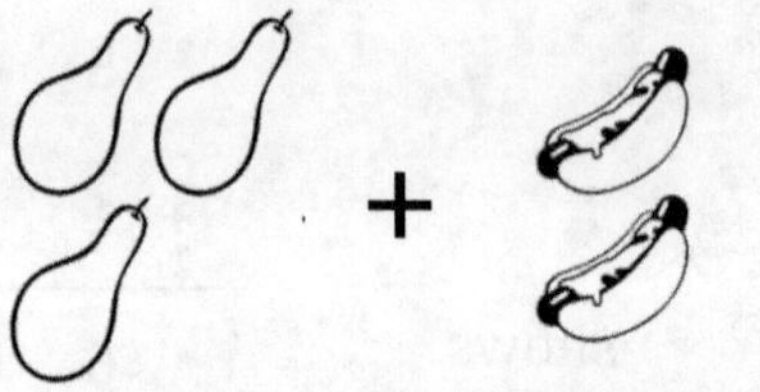

3 + 2 = 5

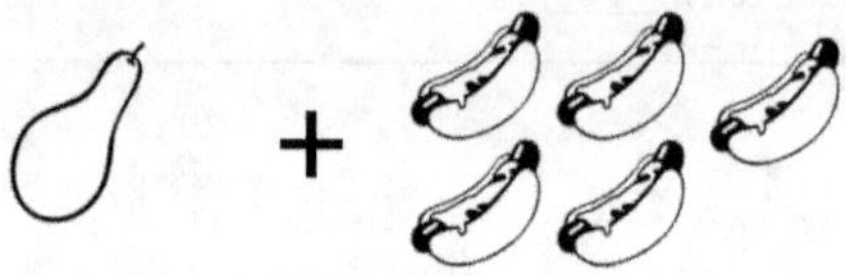

1 + 5 = 6

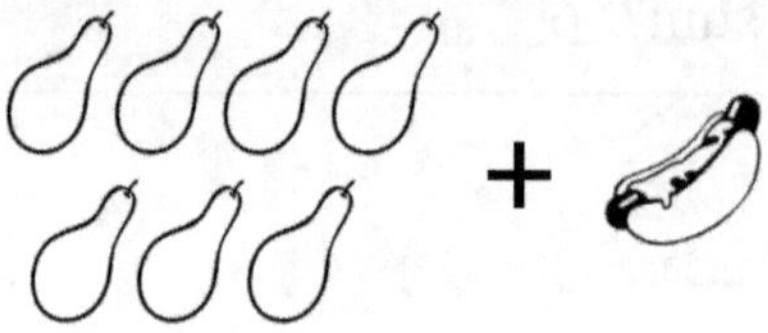

7 + 1 = 8

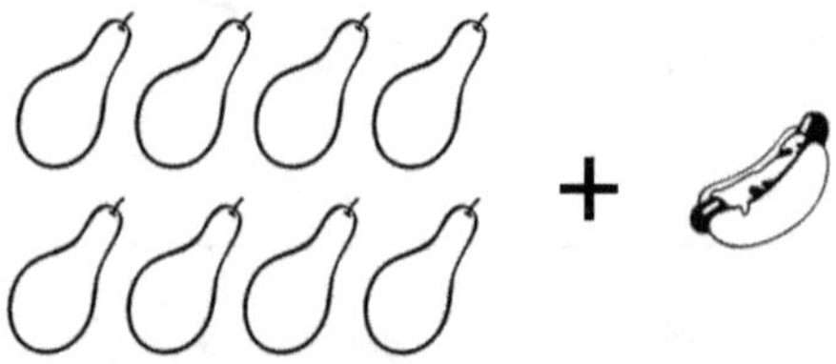

8 + 1 = 9

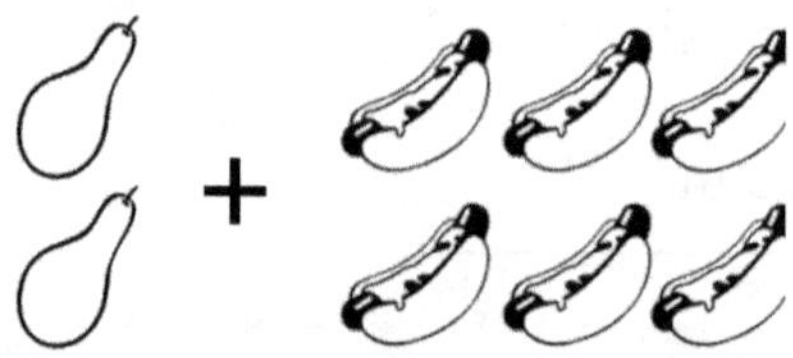

2 + 6 = 8

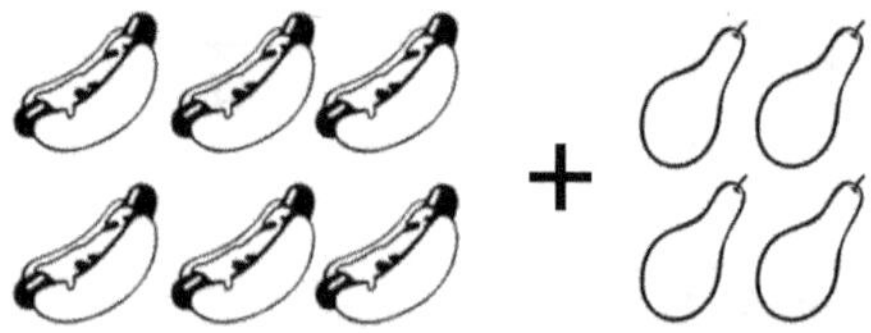

6 + 4 = 10

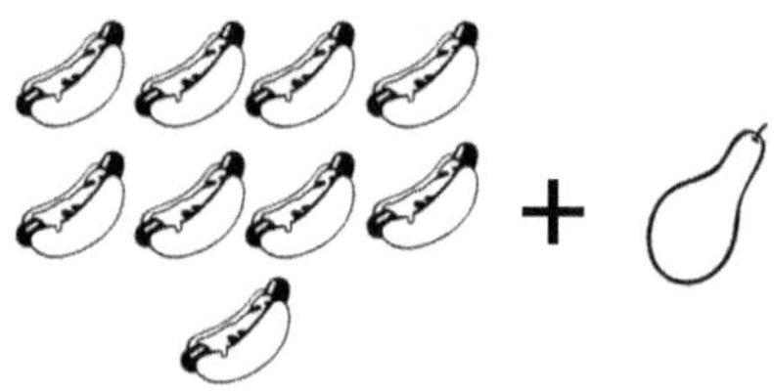

9 + 1 = 10

Copyright © Mometrix Media. You have been licensed one copy of this document for personal use only. Any other reproduction or redistribution is strictly prohibited. All rights reserved.
This content is provided for test preparation purposes only and does not imply an endorsement by Mometrix of any particular political, scientific, or religious point of view.

Day 13

Addition 1-12

  10
+ 2
= 12

  6
+ 4
= 10

  8
+ 3
= 11

  5
+ 4
= 9

  3
+ 9
= 12

  4
+ 4
= 8

  6
+ 5
= 11

  9
+ 3
= 12

Copyright © Mometrix Media. You have been licensed one copy of this document for personal use only. Any other reproduction or redistribution is strictly prohibited. All rights reserved.
This content is provided for test preparation purposes only and does not imply an endorsement by Mometrix of any particular political, scientific, or religious point of view.

DAY 14

MISSING NUMBERS ADDITION 1-10

4 + [2] = 6	[2] + 3 = 5
2 + [7] = 9	4 + [0] = 4
[3] + 2 = 5	6 + [1] = 7
5 + [4] = 9	[7] + 3 = 10
[2] + 8 = 10	2 + [6] = 8

Copyright © Mometrix Media. You have been licensed one copy of this document for personal use only. Any other reproduction or redistribution is strictly prohibited. All rights reserved.
This content is provided for test preparation purposes only and does not imply an endorsement by Mometrix of any particular political, scientific, or religious point of view.

Day 15

Missing Numbers Addition 1-20

5	+	11	=	16	14	+	5	=	19
4	+	7	=	11	12	+	5	=	17
13	+	2	=	15	6	+	14	=	20
5	+	14	=	19	10	+	3	=	13
15	+	5	=	20	2	+	16	=	18

Copyright © Mometrix Media. You have been licensed one copy of this document for personal use only. Any other reproduction or redistribution is strictly prohibited. All rights reserved.
This content is provided for test preparation purposes only and does not imply an endorsement by Mometrix of any particular political, scientific, or religious point of view.

DAY 16
ADDITION 1-20

1. 19
2. 9
3. 17
4. 12
5. 16
6. 11
7. 15
8. 3
9. 12
10. 17
11. 20
12. 10
13. 11
14. 20
15. 20
16. 9
17. 20
18. 19
19. 14
20. 20
21. 7
22. 16
23. 16
24. 12
25. 10
26. 14
27. 8
28. 20
29. 19
30. 12

Copyright © Mometrix Media. You have been licensed one copy of this document for personal use only. Any other reproduction or redistribution is strictly prohibited. All rights reserved.
This content is provided for test preparation purposes only and does not imply an endorsement by Mometrix of any particular political, scientific, or religious point of view.

Chapter 3: Subtraction

Day 17

Subtraction to 5

 -  = 0

 -  = 1

 -  = 1

 -  = 2

 -  = 4

Copyright © Mometrix Media. You have been licensed one copy of this document for personal use only. Any other reproduction or redistribution is strictly prohibited. All rights reserved.
This content is provided for test preparation purposes only and does not imply an endorsement by Mometrix of any particular political, scientific, or religious point of view.

DAY 18
WORD PROBLEMS

We see 4 

2  run away

How many are left? 2

$$\begin{array}{r} 4 \\ -\ 2 \\ \hline 2 \end{array}$$

There are 6 

2  go home

How many are left? 4

$$\begin{array}{r} 6 \\ -\ 2 \\ \hline 4 \end{array}$$

There are 6 

1  drives away

How many are left? 5

$$\begin{array}{r} 6 \\ -\ 1 \\ \hline 5 \end{array}$$

We have 4 

1  leaves

How many are left? 3

$$\begin{array}{r} 4 \\ -\ 1 \\ \hline 3 \end{array}$$

There are 7 

3  leave

How many are left? 4

$$\begin{array}{r} 7 \\ -\ 3 \\ \hline 4 \end{array}$$

There are 5 

2  drive away

What is the difference?

3

$$\begin{array}{r} 5 \\ -\ 2 \\ \hline 3 \end{array}$$

Jimmy has 2 

Sue has 1 

What is the difference? 1

$$\begin{array}{r} 2 \\ -\ 1 \\ \hline 1 \end{array}$$

Steve has 5 

He ate 1 

How many are left? 4

$$\begin{array}{r} 5 \\ -\ 1 \\ \hline 4 \end{array}$$

Copyright © Mometrix Media. You have been licensed one copy of this document for personal use only. Any other reproduction or redistribution is strictly prohibited. All rights reserved.
This content is provided for test preparation purposes only and does not imply an endorsement by Mometrix of any particular political, scientific, or religious point of view.

Day 19

Subtraction to 10

5
- 4
1

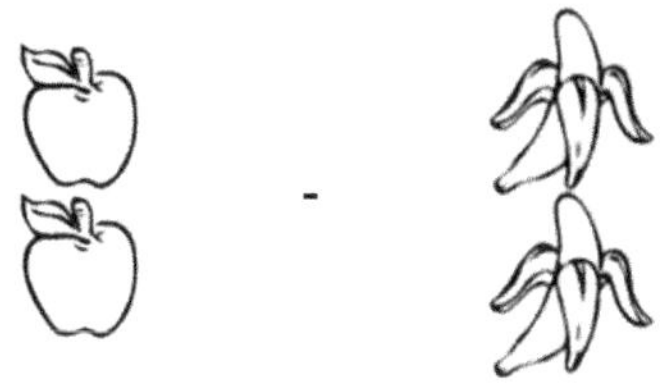

2
- 2
0

5
- 3
2

4
- 3
1

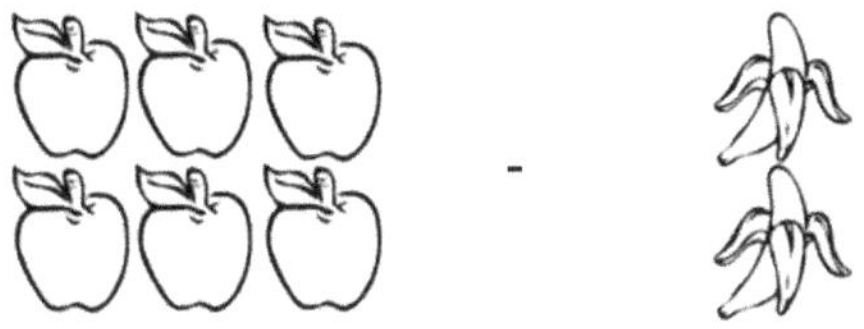

6
- 2
4

1
- 1
0

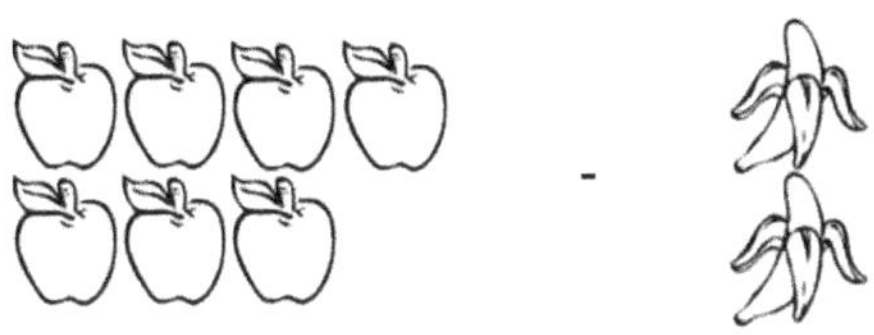

7
- 2
5

8
- 2
6

Workbook Answers

Copyright © Mometrix Media. You have been licensed one copy of this document for personal use only. Any other reproduction or redistribution is strictly prohibited. All rights reserved.
This content is provided for test preparation purposes only and does not imply an endorsement by Mometrix of any particular political, scientific, or religious point of view.

Day 20

Subtraction to 12

12
\- 2
10

10
\- 4
6

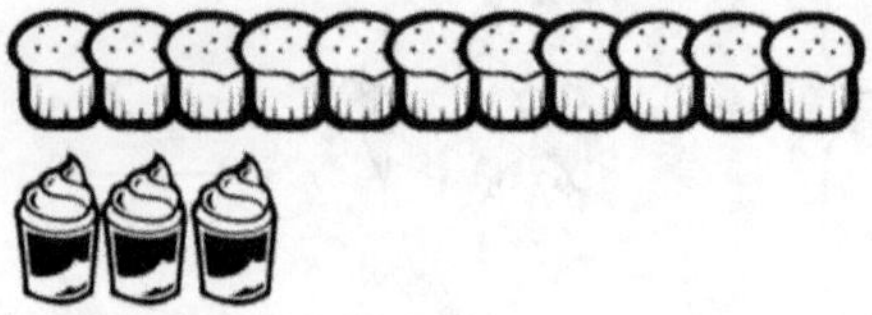

11
\- 3
8

9
\- 4
5

5
\- 4
1

7
\- 3
4

10
\- 5
5

12
\- 4
8

Copyright © Mometrix Media. You have been licensed one copy of this document for personal use only. Any other reproduction or redistribution is strictly prohibited. All rights reserved.
This content is provided for test preparation purposes only and does not imply an endorsement by Mometrix of any particular political, scientific, or religious point of view.

Day 21

Missing Numbers to 10

6	-	4	=	2
8	-	4	=	4
8	-	2	=	6
5	-	4	=	1
10	-	2	=	8

8	-	3	=	5
7	-	4	=	3
9	-	2	=	7
13	-	3	=	10
10	-	2	=	8

Copyright © Mometrix Media. You have been licensed one copy of this document for personal use only. Any other reproduction or redistribution is strictly prohibited. All rights reserved.
This content is provided for test preparation purposes only and does not imply an endorsement by Mometrix of any particular political, scientific, or religious point of view.

Day 22
Missing Numbers to 20

18	-	2	=	16
17	-	13	=	4
16	-	2	=	14
15	-	14	=	1
19	-	2	=	17

20	-	1	=	5
12	-	10	=	2
9	-	2	=	7
20	-	3	=	17
16	-	8	=	8

Copyright © Mometrix Media. You have been licensed one copy of this document for personal use only. Any other reproduction or redistribution is strictly prohibited. All rights reserved.
This content is provided for test preparation purposes only and does not imply an endorsement by Mometrix of any particular political, scientific, or religious point of view.

Day 23

Subtraction Problems to 15

1. 1
2. 5
3. 2
4. 4
5. 1
6. 9
7. 0
8. 6
9. 2
10. 1
11. 4
12. 2
13. 6
14. 0
15. 6
16. 7
17. 5
18. 3
19. 5
20. 3
21. 3
22. 1
23. 0
24. 9
25. 4
26. 4
27. 2
28. 8
29. 3
30. 6

Copyright © Mometrix Media. You have been licensed one copy of this document for personal use only. Any other reproduction or redistribution is strictly prohibited. All rights reserved.
This content is provided for test preparation purposes only and does not imply an endorsement by Mometrix of any particular political, scientific, or religious point of view.

Day 24

Subtraction Problems to 20

1. 10
2. 15
3. 12
4. 14
5. 9
6. 9
7. 1
8. 16
9. 9
10. 5
11. 19
12. 12
13. 9
14. 9
15. 13
16. 4
17. 0
18. 4
19. 12
20. 6
21. 4
22. 1
23. 10
24. 9
25. 14
26. 14
27. 17
28. 8
29. 13
30. 9

Copyright © Mometrix Media. You have been licensed one copy of this document for personal use only. Any other reproduction or redistribution is strictly prohibited. All rights reserved.
This content is provided for test preparation purposes only and does not imply an endorsement by Mometrix of any particular political, scientific, or religious point of view.

Chapter 4: Place Value, Sorting & Number Sense

Day 25

Sorting

Copyright © Mometrix Media. You have been licensed one copy of this document for personal use only. Any other reproduction or redistribution is strictly prohibited. All rights reserved.
This content is provided for test preparation purposes only and does not imply an endorsement by Mometrix of any particular political, scientific, or religious point of view.

Day 26

Grouping Addition

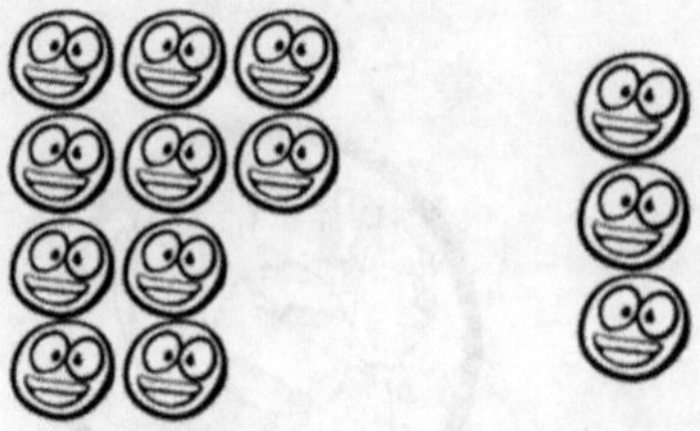

10 + 3 = 13

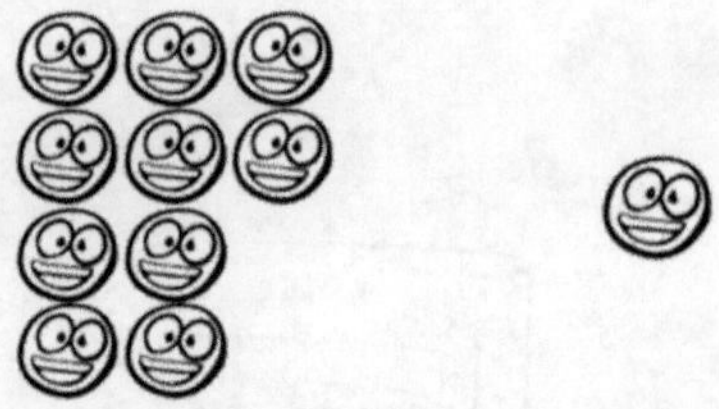

10 + 1 = 11

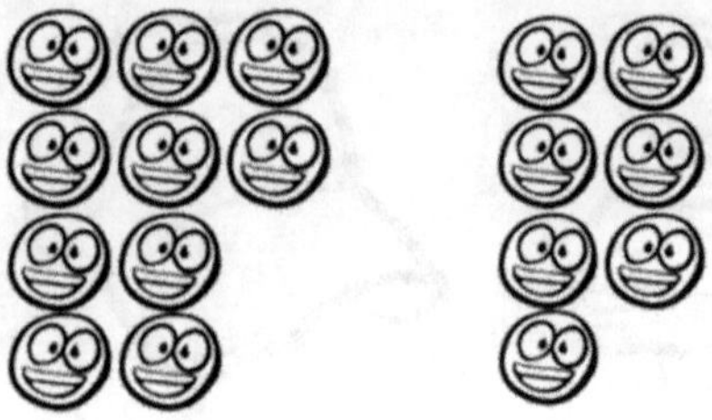

10 + 7 = 17

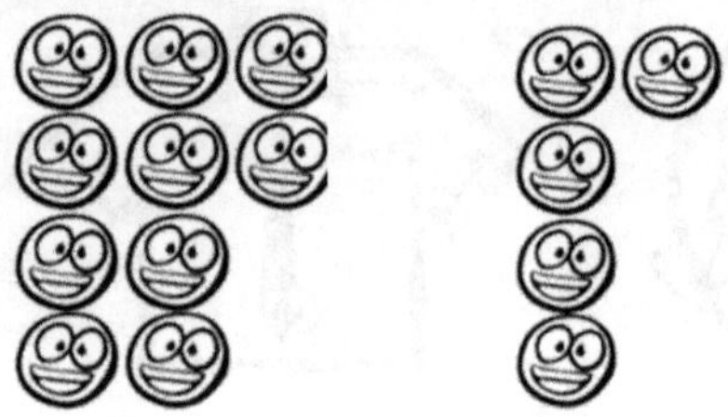

10 + 5 = 15

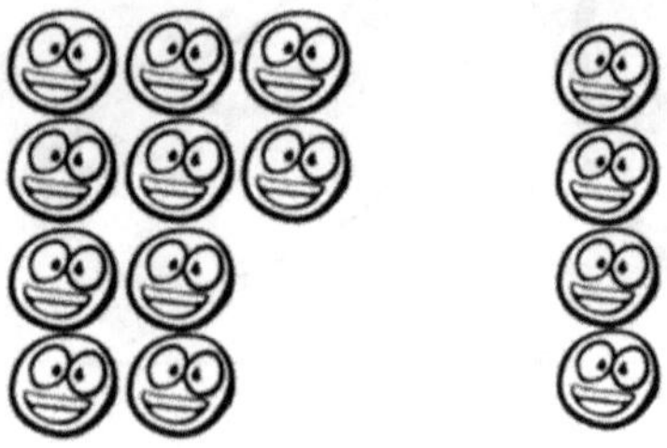

10 + 4 = 14

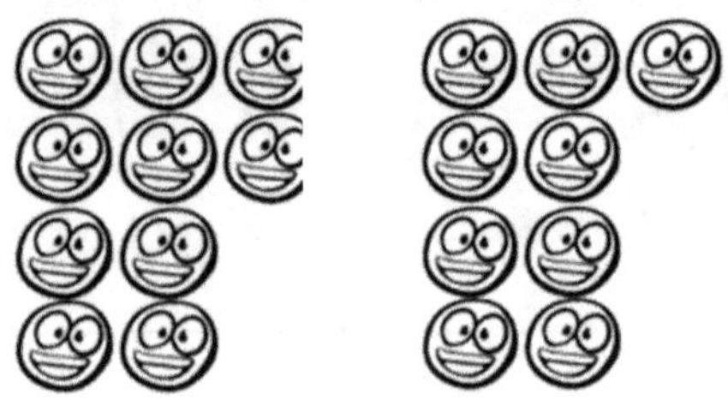

10 + 9 = 19

Copyright © Mometrix Media. You have been licensed one copy of this document for personal use only. Any other reproduction or redistribution is strictly prohibited. All rights reserved.
This content is provided for test preparation purposes only and does not imply an endorsement by Mometrix of any particular political, scientific, or religious point of view.

Day 27

Grouping Addition

 = 10  = 1

$$\begin{array}{r} 30 \\ +\ 4 \\ \hline 34 \end{array}$$

$$\begin{array}{r} 20 \\ +\ 5 \\ \hline 25 \end{array}$$

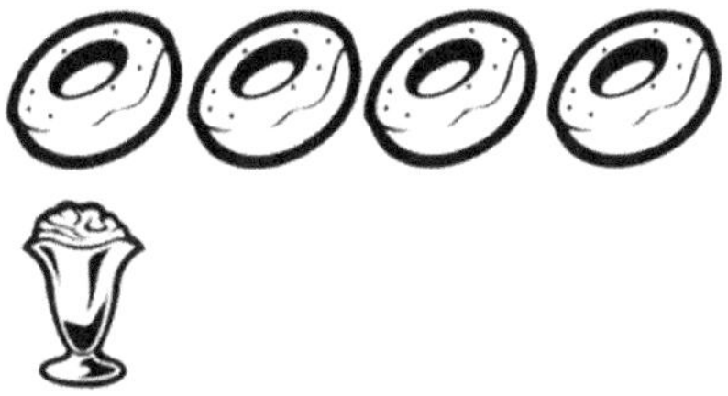

$$\begin{array}{r} 40 \\ +\ 1 \\ \hline 41 \end{array}$$

$$\begin{array}{r} 50 \\ +\ 6 \\ \hline 56 \end{array}$$

$$\begin{array}{r} 10 \\ +\ 8 \\ \hline 18 \end{array}$$

$$\begin{array}{r} 40 \\ +\ 6 \\ \hline 46 \end{array}$$

$$\begin{array}{r} 20 \\ +\ 6 \\ \hline 26 \end{array}$$

$$\begin{array}{r} 50 \\ +\ 9 \\ \hline 59 \end{array}$$

Copyright © Mometrix Media. You have been licensed one copy of this document for personal use only. Any other reproduction or redistribution is strictly prohibited. All rights reserved.
This content is provided for test preparation purposes only and does not imply an endorsement by Mometrix of any particular political, scientific, or religious point of view.

Day 28

Grouping Subtraction

 = 10  = 1

$$\begin{array}{r} 40 \\ -\ 2 \\ \hline 38 \end{array}$$

$$\begin{array}{r} 50 \\ -\ 4 \\ \hline 46 \end{array}$$

$$\begin{array}{r} 20 \\ -\ 6 \\ \hline 14 \end{array}$$

$$\begin{array}{r} 60 \\ -\ 1 \\ \hline 59 \end{array}$$

$$\begin{array}{r} 10 \\ -\ 3 \\ \hline 7 \end{array}$$

$$\begin{array}{r} 50 \\ -\ 5 \\ \hline 45 \end{array}$$

$$\begin{array}{r} 70 \\ -10 \\ \hline 60 \end{array}$$

$$\begin{array}{r} 30 \\ -\ 6 \\ \hline 24 \end{array}$$

Copyright © Mometrix Media. You have been licensed one copy of this document for personal use only. Any other reproduction or redistribution is strictly prohibited. All rights reserved.
This content is provided for test preparation purposes only and does not imply an endorsement by Mometrix of any particular political, scientific, or religious point of view.

Day 29

Comparing Groups

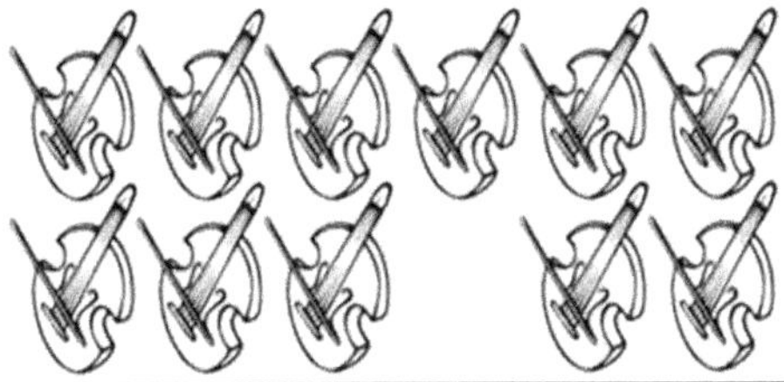

<

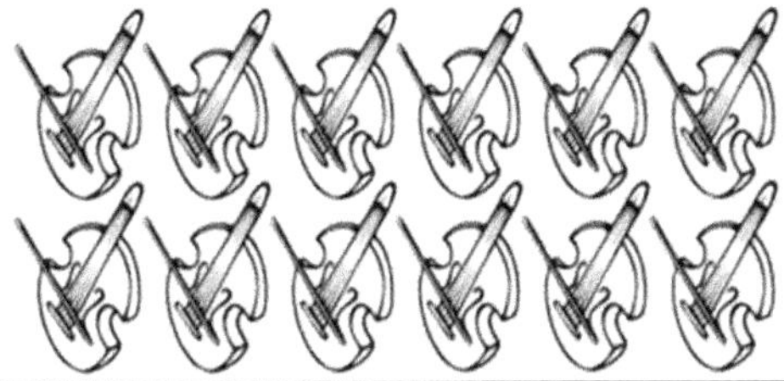

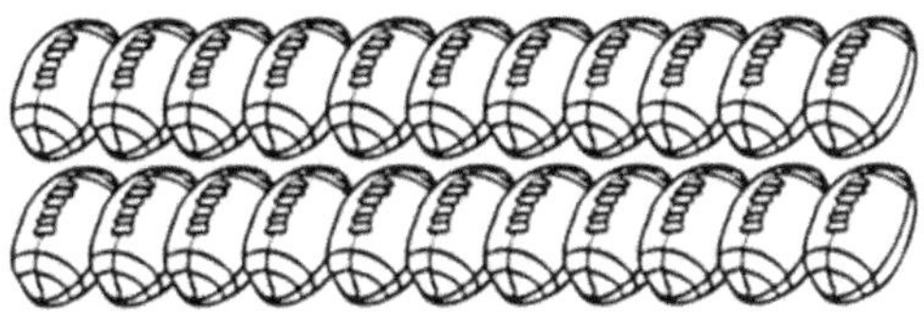

>

<

<

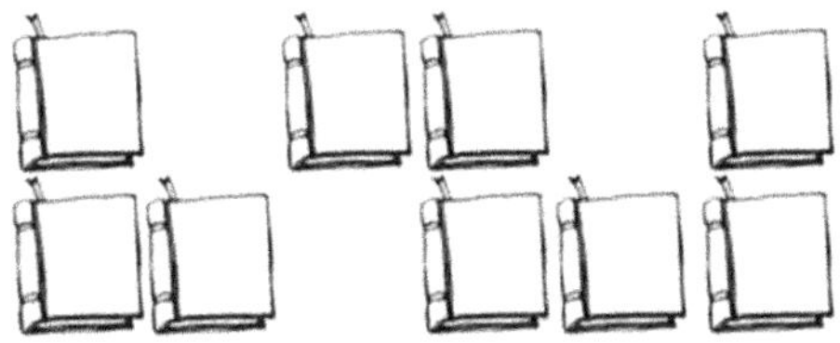

>

Copyright © Mometrix Media. You have been licensed one copy of this document for personal use only. Any other reproduction or redistribution is strictly prohibited. All rights reserved.
This content is provided for test preparation purposes only and does not imply an endorsement by Mometrix of any particular political, scientific, or religious point of view.

DAY 30
COMPARING GROUPS

>

<

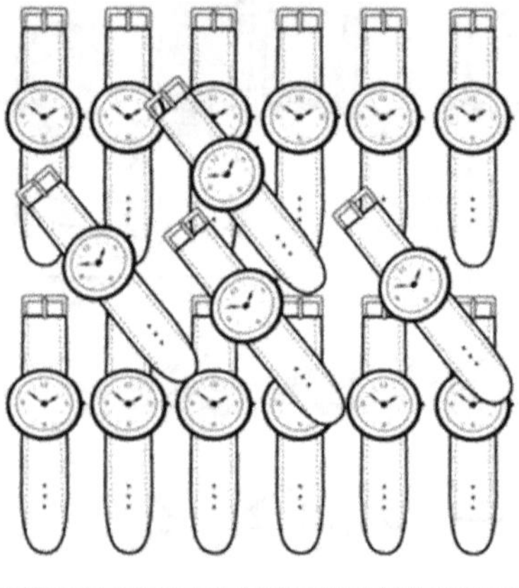

>

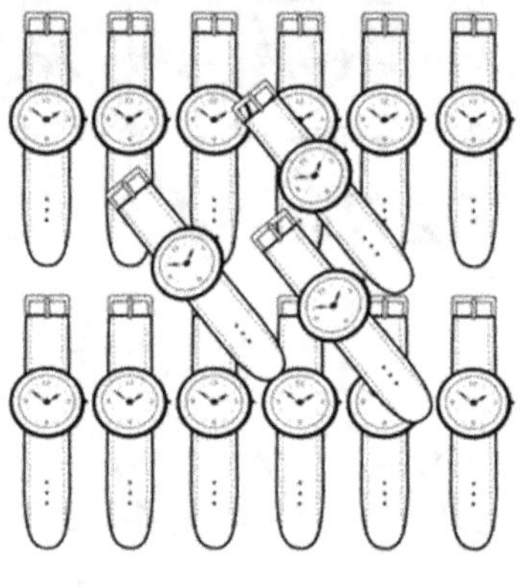

Copyright © Mometrix Media. You have been licensed one copy of this document for personal use only. Any other reproduction or redistribution is strictly prohibited. All rights reserved.
This content is provided for test preparation purposes only and does not imply an endorsement by Mometrix of any particular political, scientific, or religious point of view.

Chapter 5: Time

Day 31

Identifying Time

9:00

6:00

8:00

2:00

3:30

8:30

5:30

2:30

7:30

10:30

Copyright © Mometrix Media. You have been licensed one copy of this document for personal use only. Any other reproduction or redistribution is strictly prohibited. All rights reserved.
This content is provided for test preparation purposes only and does not imply an endorsement by Mometrix of any particular political, scientific, or religious point of view.

Day 32

Writing Time

Copyright © Mometrix Media. You have been licensed one copy of this document for personal use only. Any other reproduction or redistribution is strictly prohibited. All rights reserved.
This content is provided for test preparation purposes only and does not imply an endorsement by Mometrix of any particular political, scientific, or religious point of view.

Copyright © Mometrix Media. You have been licensed one copy of this document for personal use only. Any other reproduction or redistribution is strictly prohibited. All rights reserved.
This content is provided for test preparation purposes only and does not imply an endorsement by Mometrix of any particular political, scientific, or religious point of view.

Day 33
Drawing Time

Copyright © Mometrix Media. You have been licensed one copy of this document for personal use only. Any other reproduction or redistribution is strictly prohibited. All rights reserved.
This content is provided for test preparation purposes only and does not imply an endorsement by Mometrix of any particular political, scientific, or religious point of view.

Copyright © Mometrix Media. You have been licensed one copy of this document for personal use only. Any other reproduction or redistribution is strictly prohibited. All rights reserved.
This content is provided for test preparation purposes only and does not imply an endorsement by Mometrix of any particular political, scientific, or religious point of view.

DAY 34
TIME-MATCHING

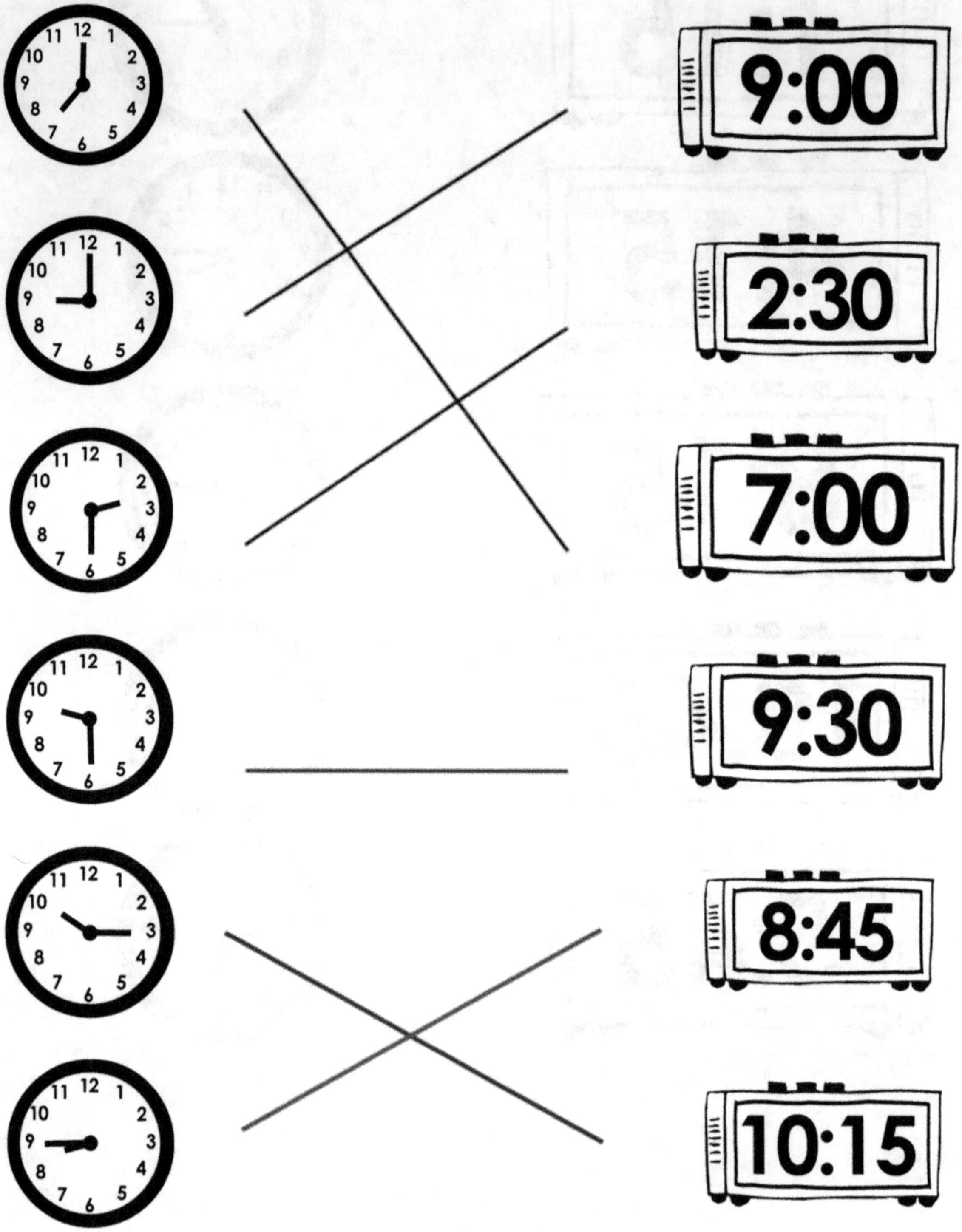

Copyright © Mometrix Media. You have been licensed one copy of this document for personal use only. Any other reproduction or redistribution is strictly prohibited. All rights reserved.
This content is provided for test preparation purposes only and does not imply an endorsement by Mometrix of any particular political, scientific, or religious point of view.

Chapter 6: Measurements

DAY 35

IDENTIFYING LARGER OBJECTS

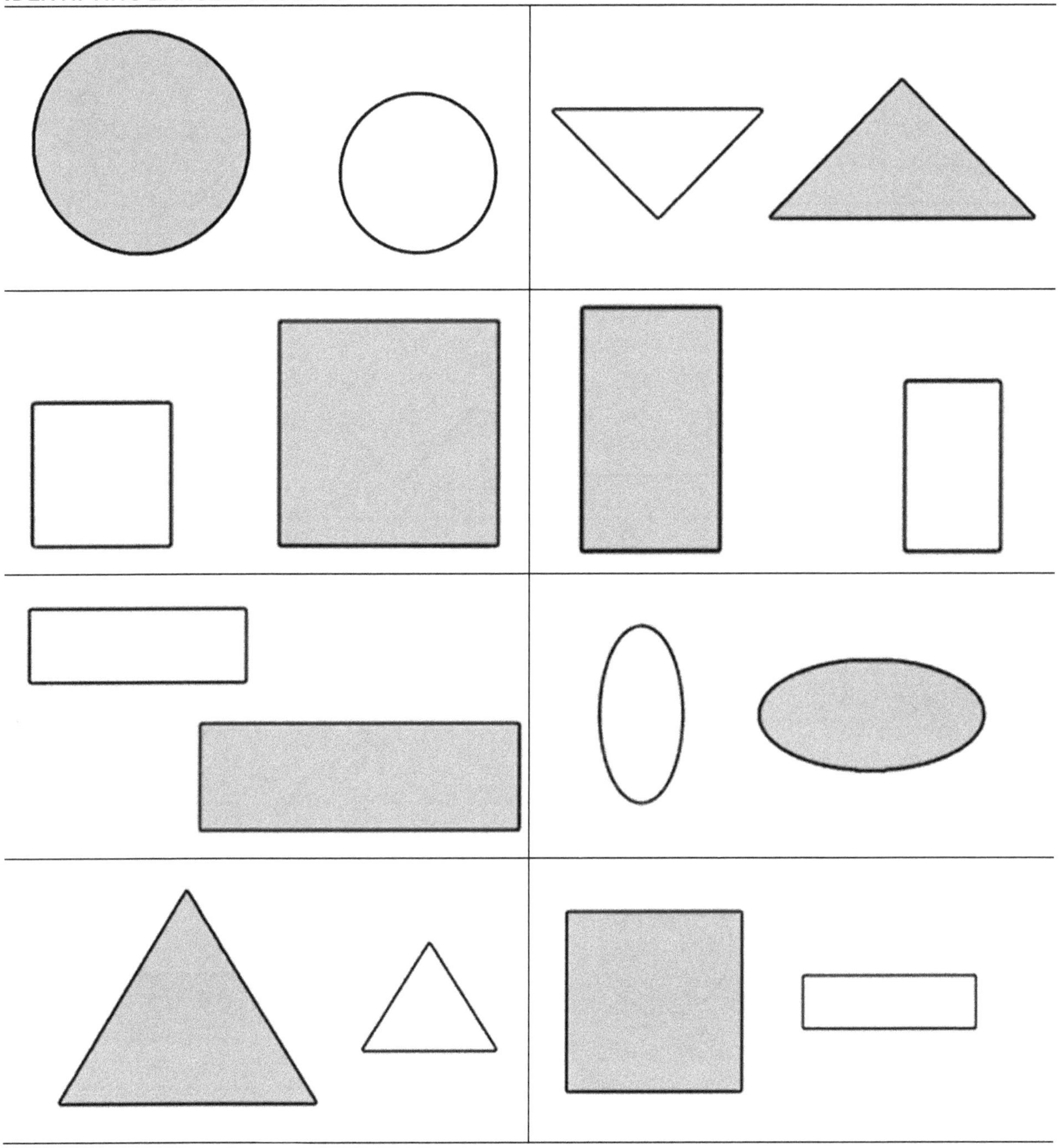

Copyright © Mometrix Media. You have been licensed one copy of this document for personal use only. Any other reproduction or redistribution is strictly prohibited. All rights reserved.
This content is provided for test preparation purposes only and does not imply an endorsement by Mometrix of any particular political, scientific, or religious point of view.

DAY 36

IDENTIFYING SMALLER OBJECTS

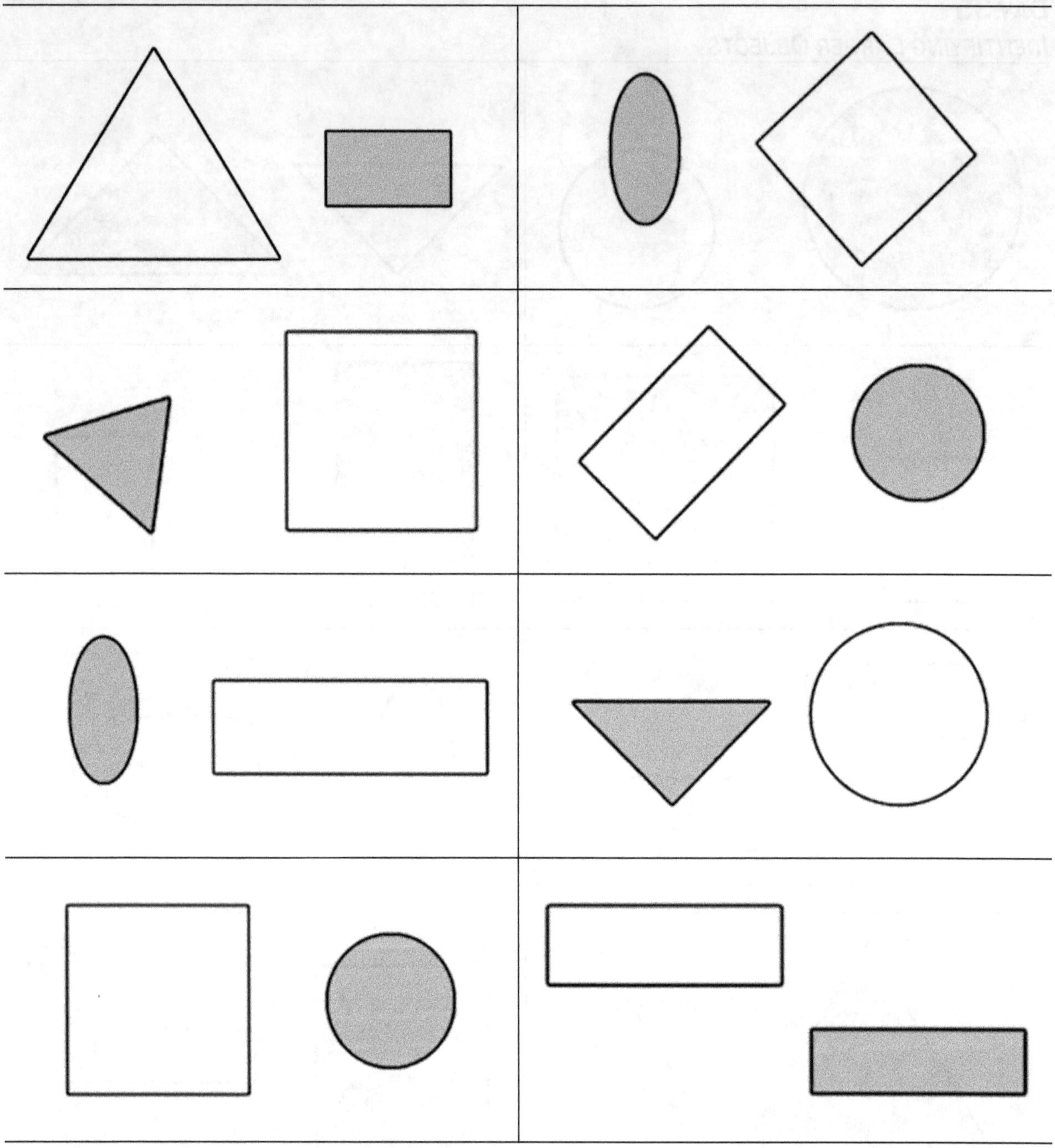

Copyright © Mometrix Media. You have been licensed one copy of this document for personal use only. Any other reproduction or redistribution is strictly prohibited. All rights reserved.
This content is provided for test preparation purposes only and does not imply an endorsement by Mometrix of any particular political, scientific, or religious point of view.

Day 37
Measuring Units: Length

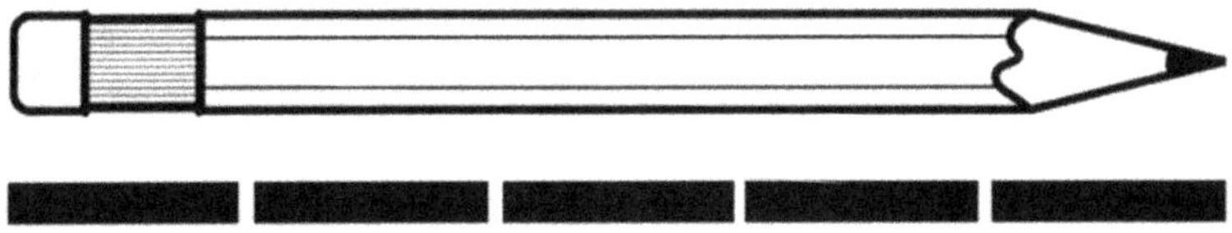

5
Units

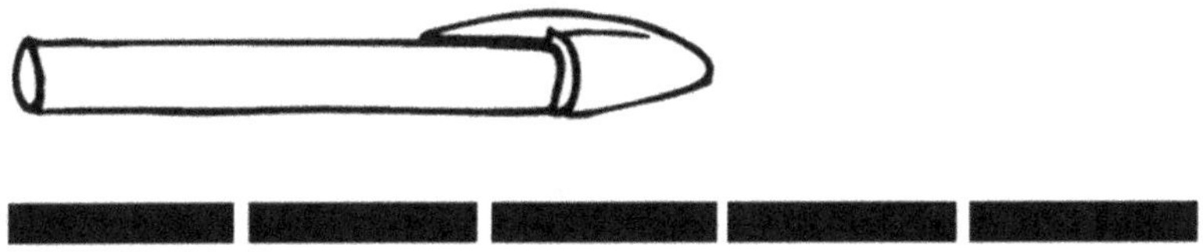

3
Units

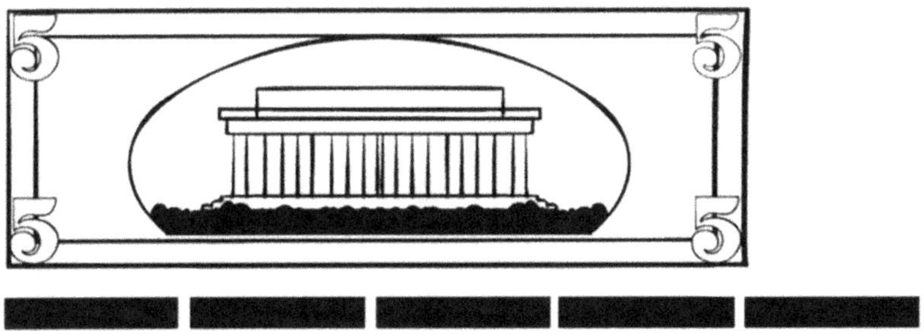

4
Units

2
Units

Copyright © Mometrix Media. You have been licensed one copy of this document for personal use only. Any other reproduction or redistribution is strictly prohibited. All rights reserved.
This content is provided for test preparation purposes only and does not imply an endorsement by Mometrix of any particular political, scientific, or religious point of view.

Day 38

Measuring Units: Height

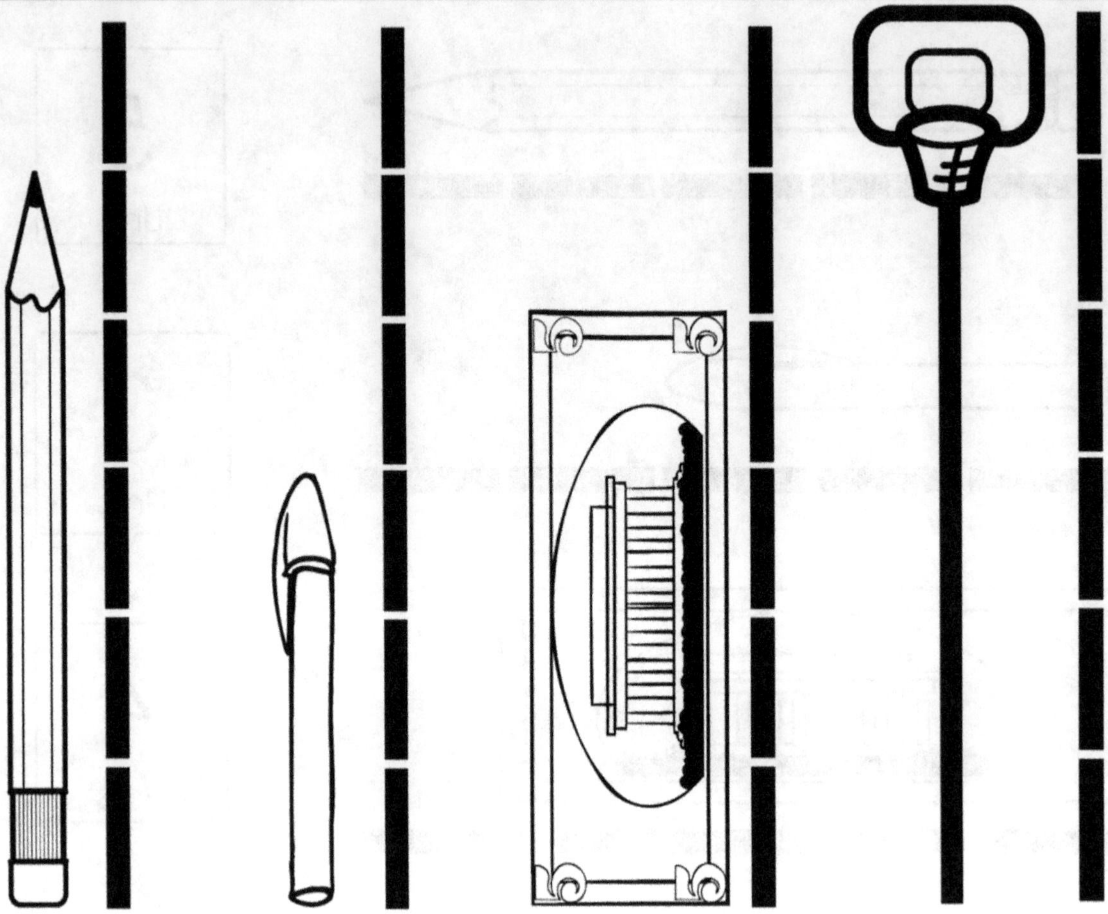

5 Units	3 Units	4 Units	6 Units

Copyright © Mometrix Media. You have been licensed one copy of this document for personal use only. Any other reproduction or redistribution is strictly prohibited. All rights reserved.
This content is provided for test preparation purposes only and does not imply an endorsement by Mometrix of any particular political, scientific, or religious point of view.

DAY 39
MEASURING INCHES

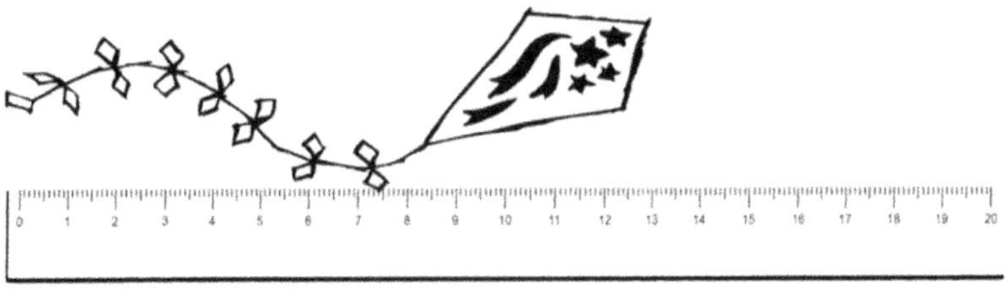

13 inches

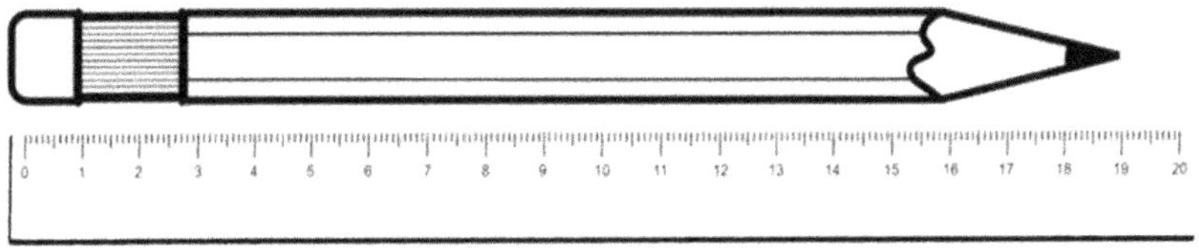

19 inches

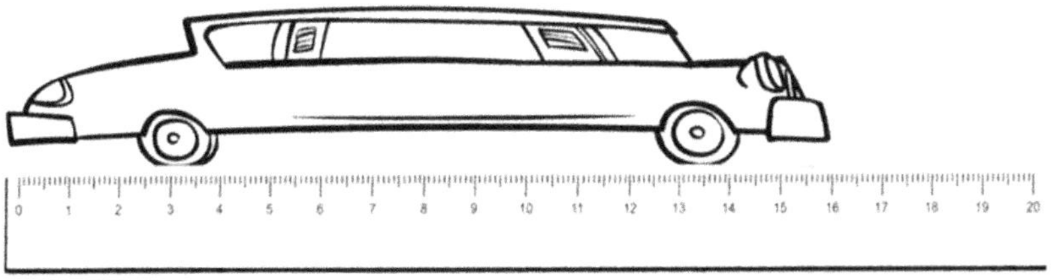

16 inches

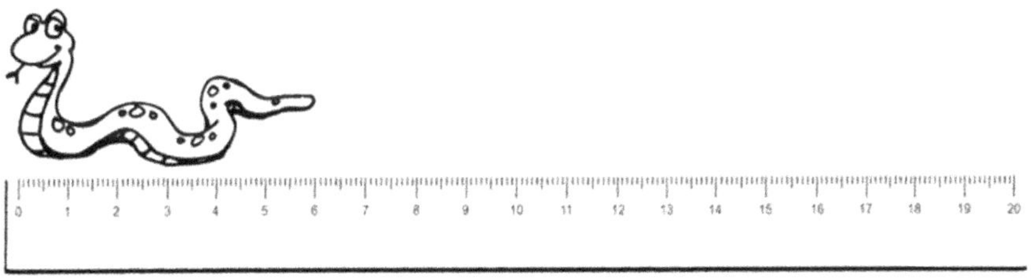

6 inches

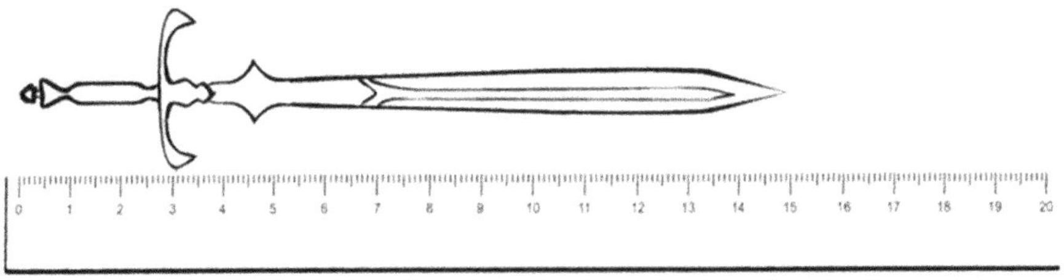

15 inches

Copyright © Mometrix Media. You have been licensed one copy of this document for personal use only. Any other reproduction or redistribution is strictly prohibited. All rights reserved.
This content is provided for test preparation purposes only and does not imply an endorsement by Mometrix of any particular political, scientific, or religious point of view.

Day 40
Measuring Half-Inches

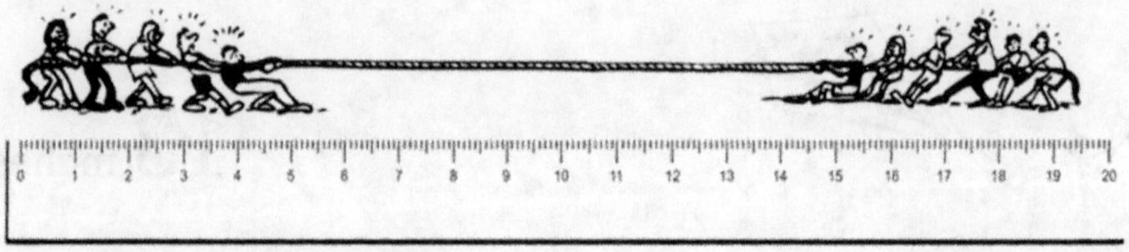

$19\frac{1}{2}$ inches

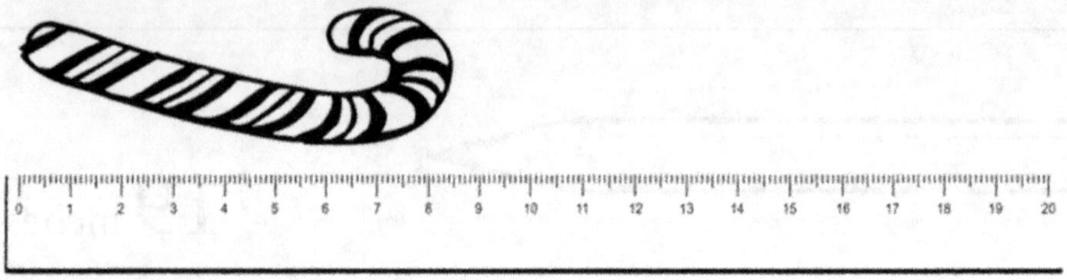

$8\frac{1}{2}$ inches

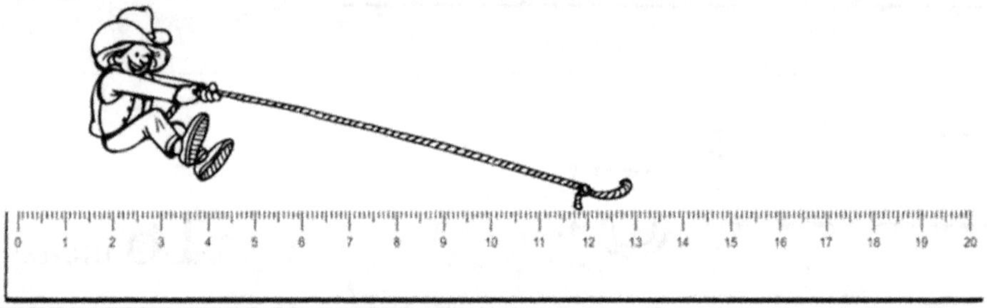

$11\frac{1}{2}$ inches

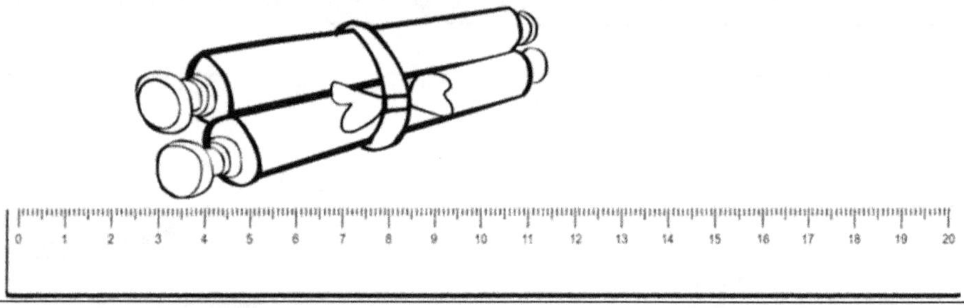

9 inches

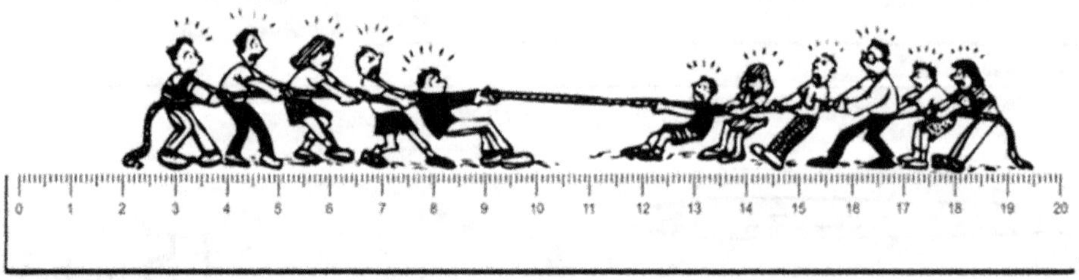

$17\frac{1}{2}$ inches

Copyright © Mometrix Media. You have been licensed one copy of this document for personal use only. Any other reproduction or redistribution is strictly prohibited. All rights reserved.
This content is provided for test preparation purposes only and does not imply an endorsement by Mometrix of any particular political, scientific, or religious point of view.

Day 41
Measuring Height

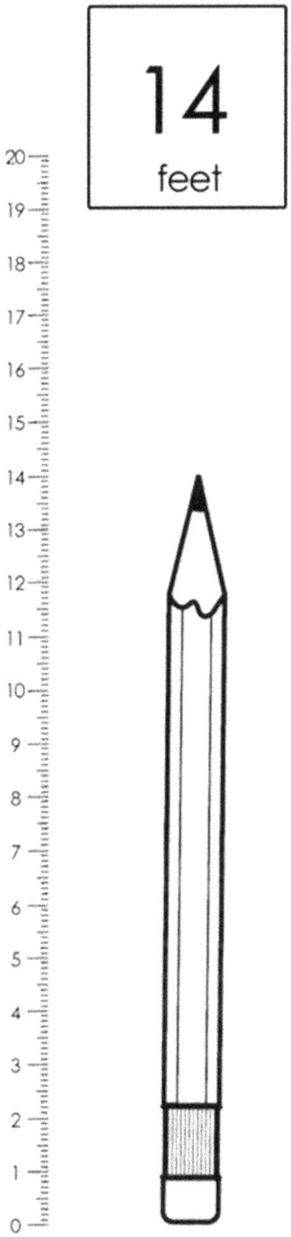

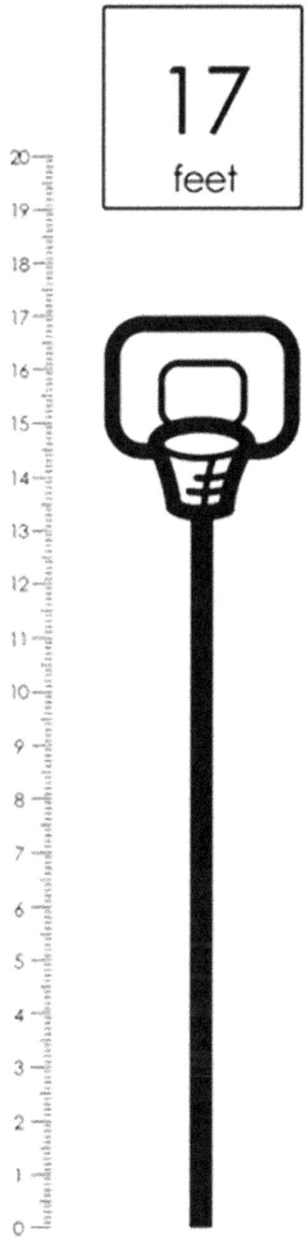

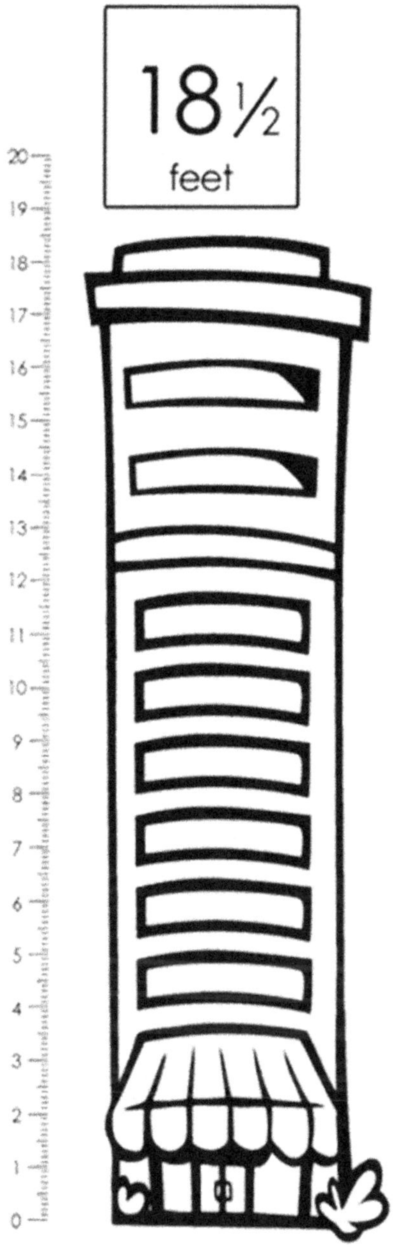

Copyright © Mometrix Media. You have been licensed one copy of this document for personal use only. Any other reproduction or redistribution is strictly prohibited. All rights reserved.
This content is provided for test preparation purposes only and does not imply an endorsement by Mometrix of any particular political, scientific, or religious point of view.

Day 42

Identifying Object Weights

Copyright © Mometrix Media. You have been licensed one copy of this document for personal use only. Any other reproduction or redistribution is strictly prohibited. All rights reserved.
This content is provided for test preparation purposes only and does not imply an endorsement by Mometrix of any particular political, scientific, or religious point of view.

Day 43

Writing Object Weights

3 Pounds

1 Pounds

7 Pounds

5 Pounds

2 Pounds

9 Pounds

Copyright © Mometrix Media. You have been licensed one copy of this document for personal use only. Any other reproduction or redistribution is strictly prohibited. All rights reserved.
This content is provided for test preparation purposes only and does not imply an endorsement by Mometrix of any particular political, scientific, or religious point of view.

Day 44

Writing Temperature

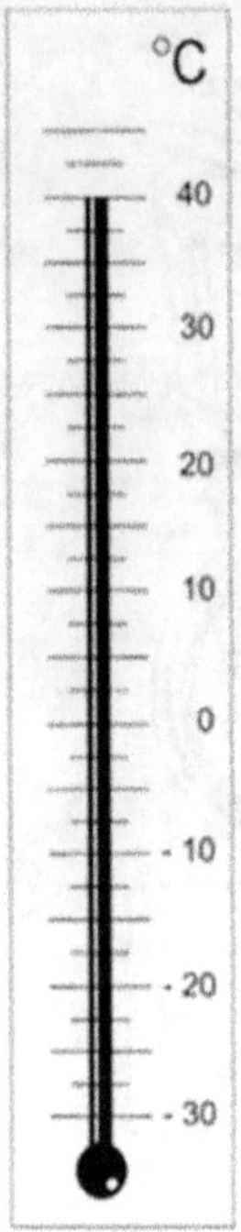

40°

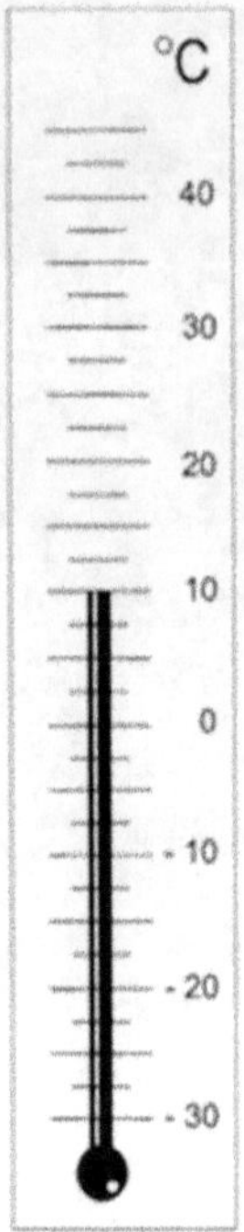

10°

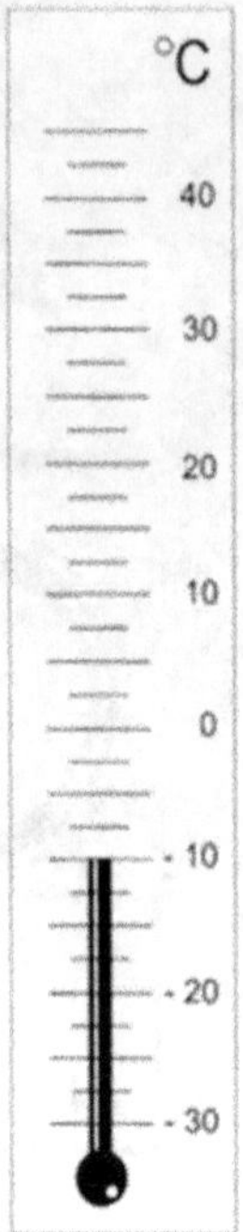

-10°

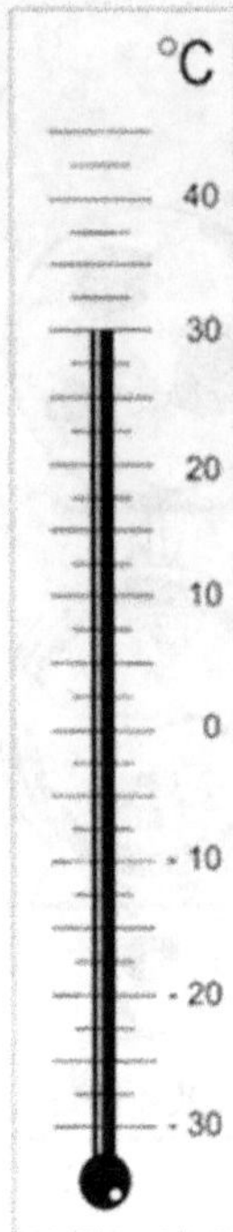

30°

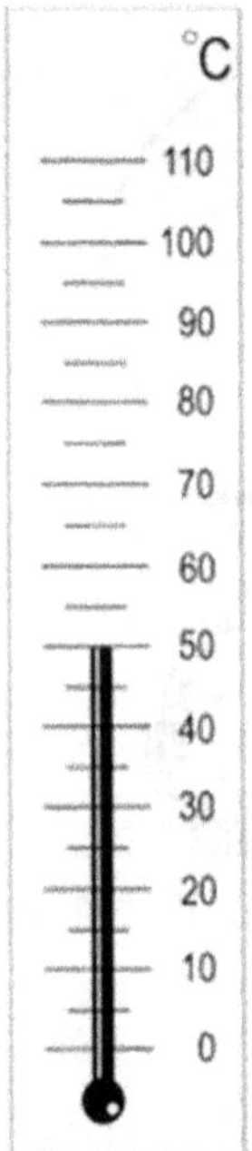

50°

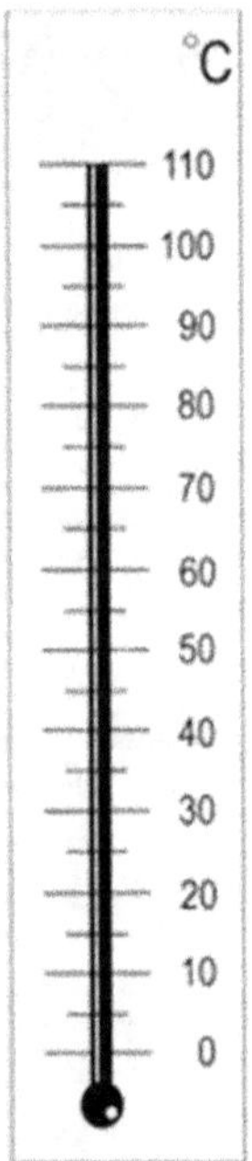

110°

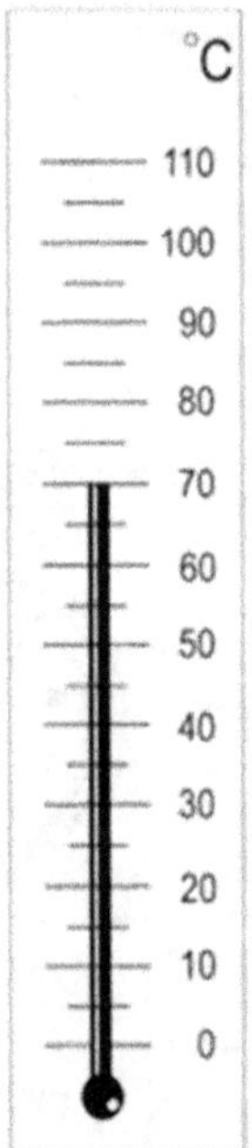

70°

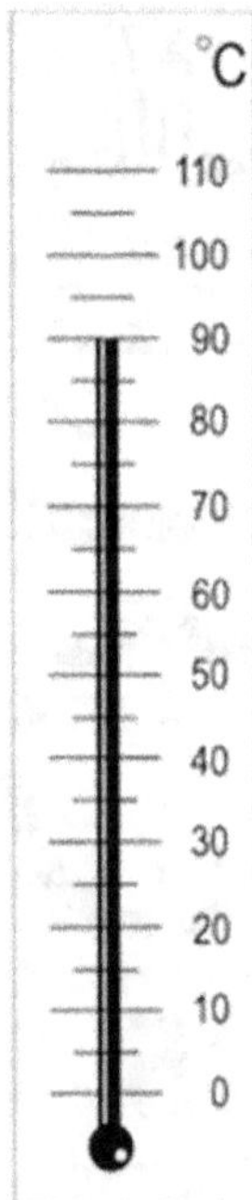

90°

Copyright © Mometrix Media. You have been licensed one copy of this document for personal use only. Any other reproduction or redistribution is strictly prohibited. All rights reserved.
This content is provided for test preparation purposes only and does not imply an endorsement by Mometrix of any particular political, scientific, or religious point of view.

Chapter 7: Geometry

Day 45

Writing Shapes

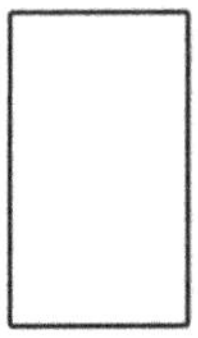	Rectangle	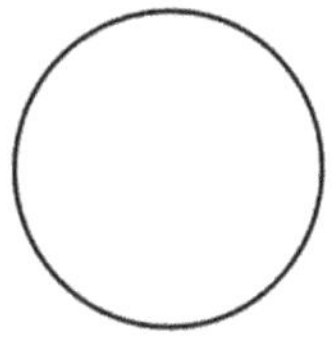	Circle
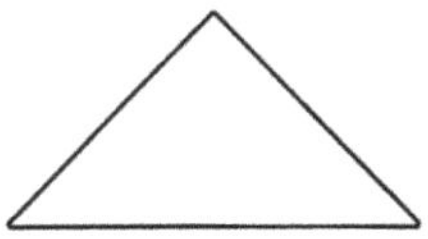	Triangle		Square
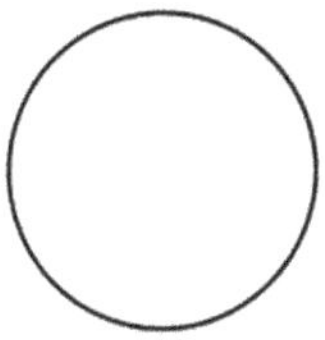	Circle	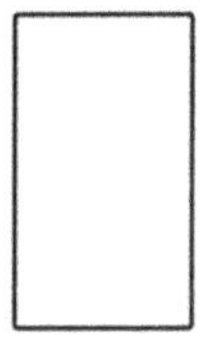	Rectangle
	Square	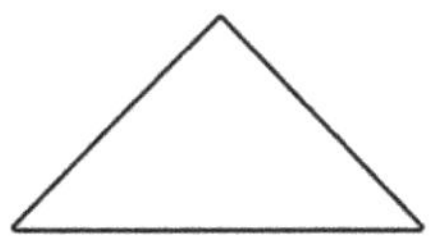	Triangle

Copyright © Mometrix Media. You have been licensed one copy of this document for personal use only. Any other reproduction or redistribution is strictly prohibited. All rights reserved.
This content is provided for test preparation purposes only and does not imply an endorsement by Mometrix of any particular political, scientific, or religious point of view.

DAY 46
DRAWING SHAPES

Square		Triangle	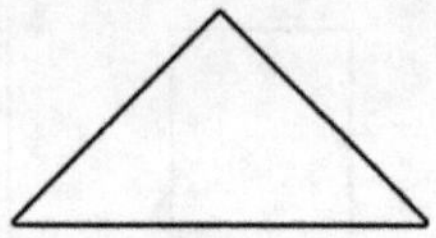
Circle	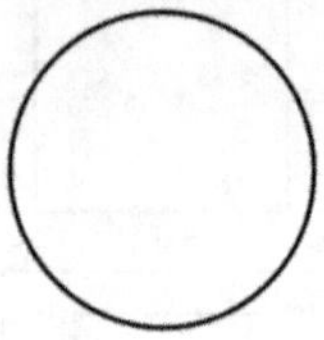	Rectangle	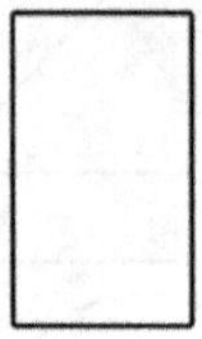
Triangle	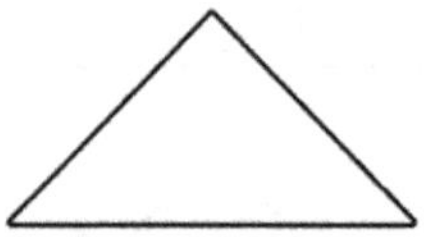	Circle	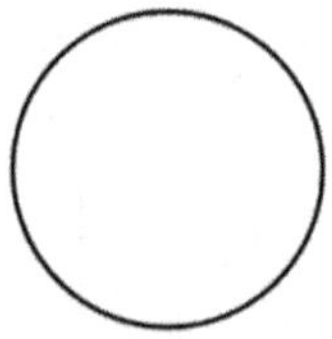
Square		Rectangle	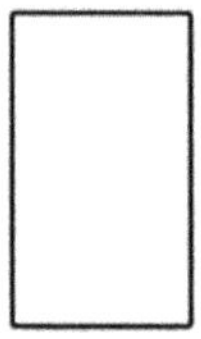

Copyright © Mometrix Media. You have been licensed one copy of this document for personal use only. Any other reproduction or redistribution is strictly prohibited. All rights reserved.
This content is provided for test preparation purposes only and does not imply an endorsement by Mometrix of any particular political, scientific, or religious point of view.

DAY 47

COUNTING SIDES OF SHAPES

Shape	Answer	Shape	Answer
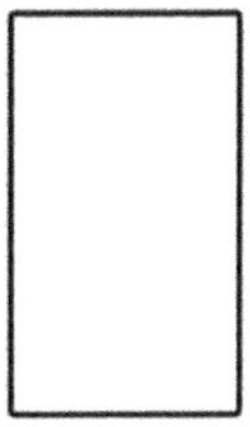	4	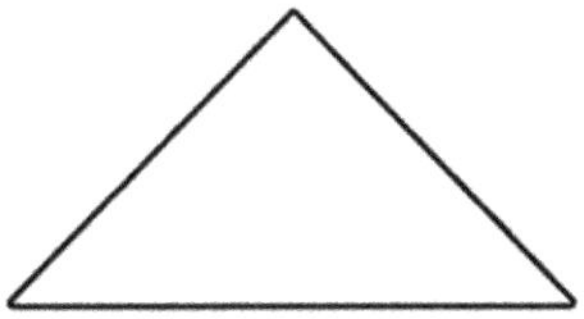	3
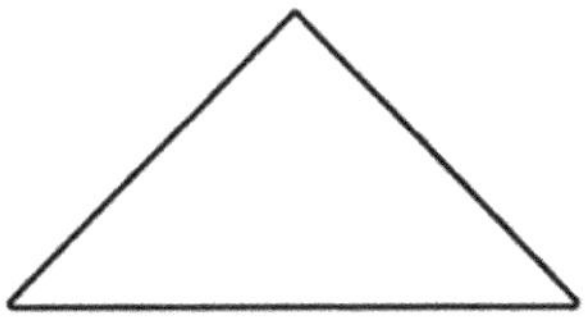	3		4
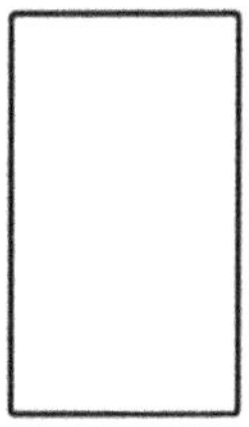	4	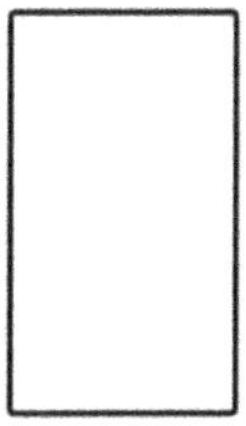	4
	4	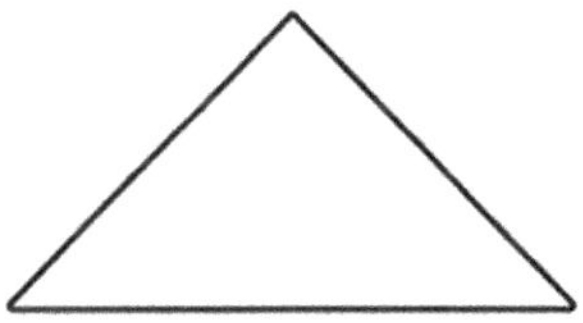	3
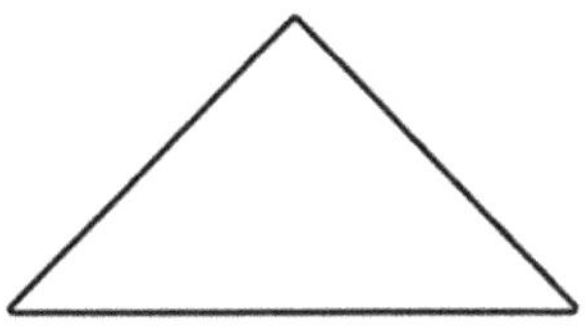	4	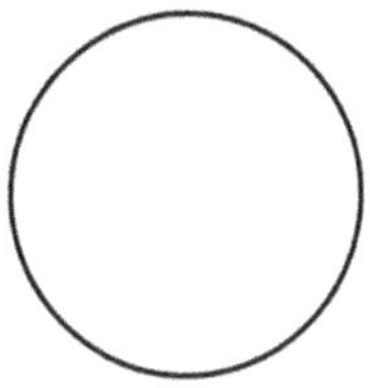	1

Copyright © Mometrix Media. You have been licensed one copy of this document for personal use only. Any other reproduction or redistribution is strictly prohibited. All rights reserved.
This content is provided for test preparation purposes only and does not imply an endorsement by Mometrix of any particular political, scientific, or religious point of view.

DAY 48
WRITING ADVANCED SHAPES

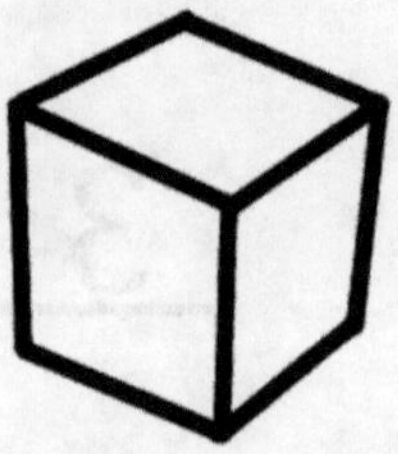 Cube	 Sphere
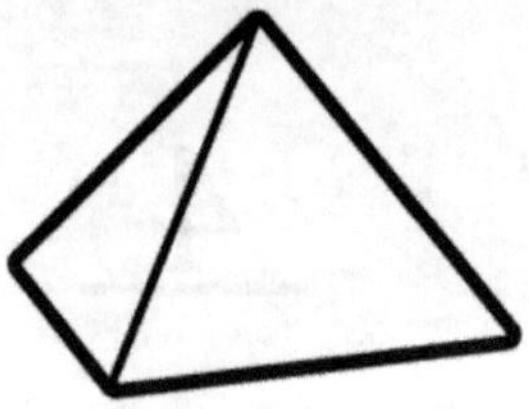 Pyramid	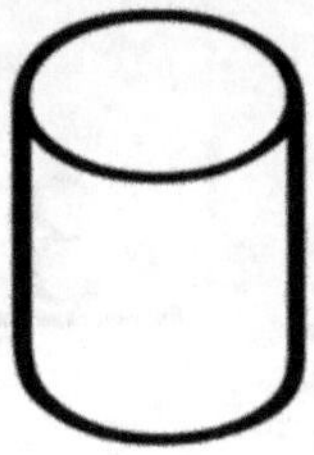 Cylinder
 Sphere	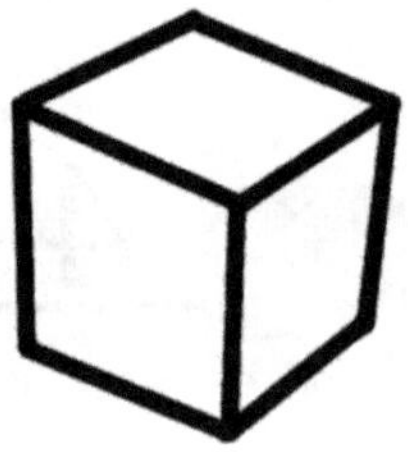 Cube
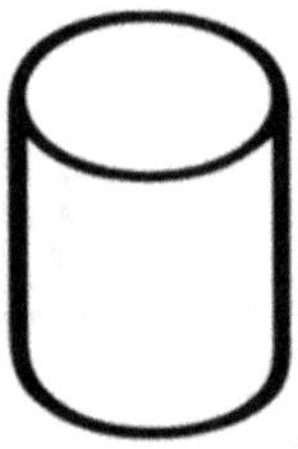 Cylinder	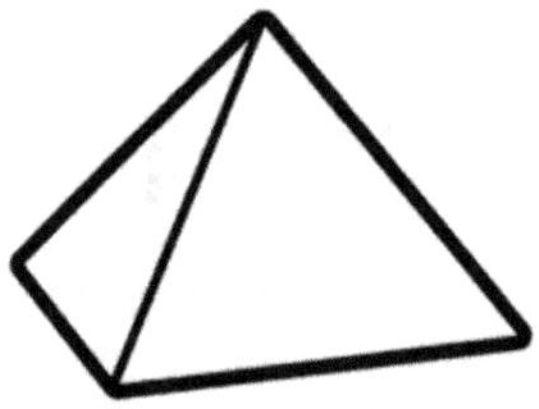 Pyramid

Copyright © Mometrix Media. You have been licensed one copy of this document for personal use only. Any other reproduction or redistribution is strictly prohibited. All rights reserved.
This content is provided for test preparation purposes only and does not imply an endorsement by Mometrix of any particular political, scientific, or religious point of view.

Day 49
Counting & Coloring Shapes

Color all the rectangles red. How many rectangles did you count? 7

Color all the circles blue. How many circles did you count? 5

Color all the squares green. How many squares did you count? 5

Color all the triangles orange. How many triangles did you count? 6

Copyright © Mometrix Media. You have been licensed one copy of this document for personal use only. Any other reproduction or redistribution is strictly prohibited. All rights reserved.
This content is provided for test preparation purposes only and does not imply an endorsement by Mometrix of any particular political, scientific, or religious point of view.

Day 50

Counting & Coloring Advanced Shapes

Color all the spheres red. How many spheres did you count? 5

Color all the cubes blue. How many cubes did you count? 4

Color all the cylinders green. How many cylinders did you count? 4

Color all the pyramids orange. How many pyramids did you count? 5

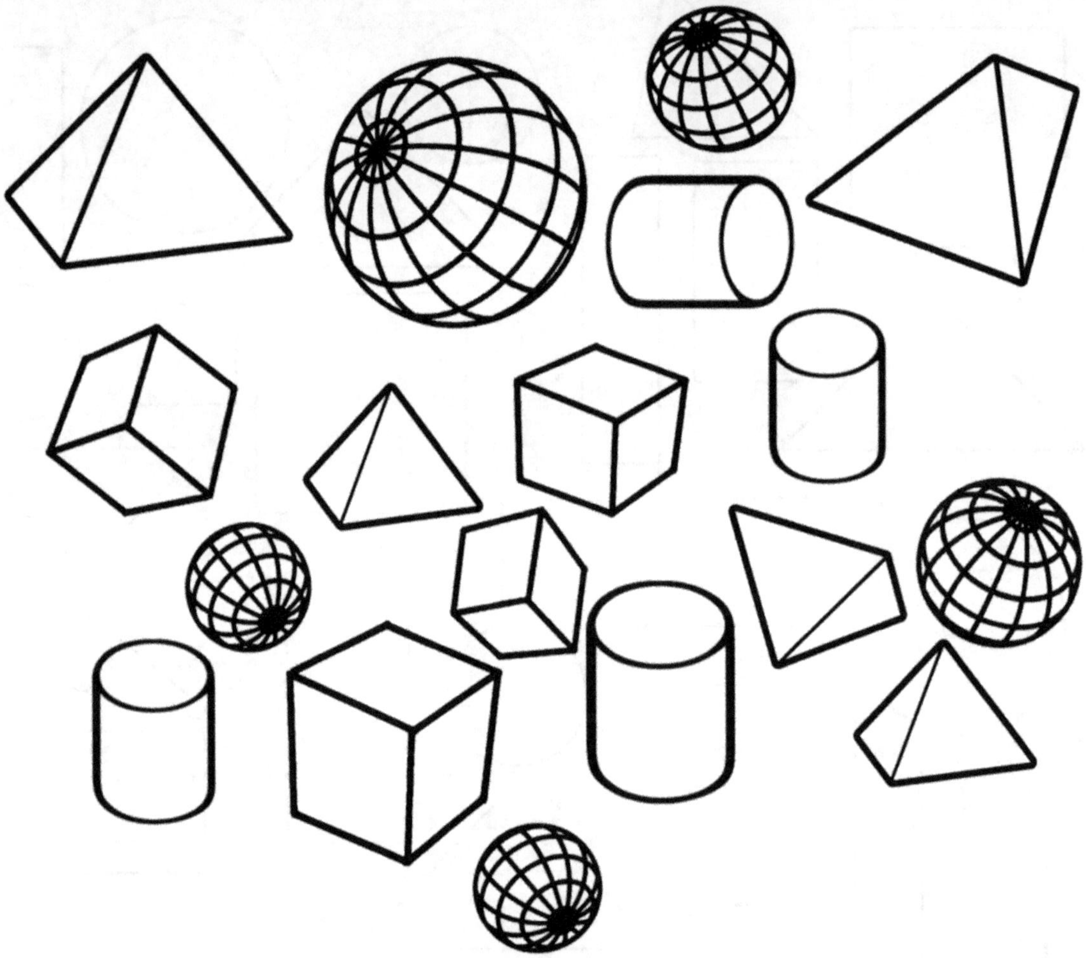

Copyright © Mometrix Media. You have been licensed one copy of this document for personal use only. Any other reproduction or redistribution is strictly prohibited. All rights reserved.
This content is provided for test preparation purposes only and does not imply an endorsement by Mometrix of any particular political, scientific, or religious point of view.

www.ingramcontent.com/pod-product-compliance
Lightning Source LLC
LaVergne TN
LVHW082353200926
848481LV00012B/3550